Collector's Guide to
Children's Automobiles

G.G. Weiner

Schiffer Publishing Ltd
4880 Lower Valley Road, Atglen, PA 19310 USA

Left: Eureka, France, circa 1927 Sport Model No. 1/27, Code EK 05*. Light gauge steel panels bolted onto twin steel runners. Chrome radiator with working headlights fed by six-volt battery under the hood. Working handbrake to rear offside wheel. Balloon pressed shteel wheels with inserted solid rubber tires. Authentic pressed tin dashboard clocks. Adjustable plain (not upholstered) steel seat.

Right: Eureka, France circa 1928 Bugatti Biplace (tandem). Heavy gauge pressed steel panels bolted onto a steel box section chassis with wooden running boards, chrome radiator with dummy headlights, radiator cap and enamel Eureka badge, chrome steering wheel and rear grab handle. Polished pierced alloy wheels with pneumatic treaded tires, authentic pressed tin dashboard clocks on wood facia. Re-upholstered black leather seats, re-painted in French racing blue. A superb restoration. Included is a factory option spare wheel and leather bonnet strap (impress date stamped 1929) Note, this does not appear to have ever been fitted with double bumpers or the handbrake normally located to right (inside) of the front cab, or a bulb-horn. Ref: Grandvoinnet M. Eureka (see pages 46-49.)
E. R. Fisher, U.K. £3,500 - £4,000.

Library of Congress Card Number: 00-111509

Designed by John P. Cheek
Type set in Americana XBd BT/Souvenir Lt BT

ISBN: 0-7643-1239-1
Printed in China
1 2 3 4

Published by Schiffer Publishing Ltd.
4880 Lower Valley Road
Atglen, PA 19310
Phone: (610) 593-1777; Fax: (610) 593-2002
E-mail: Schifferbk@aol.com
Please visit our web site catalog at www.schifferbooks.com or write for a free catalog.

We are always looking for authors to write books on new and related subjects. If you have an idea for a book, please contact us at the above address.

This book may be purchased from the publisher.
Please include $3.95 for shipping.

In Europe, Schiffer books are distributed by
Bushwood Books
6 Marksbury Ave.
Kew Gardens
Surrey TW9 4JF England
Phone: 44 (0)20-8392-8585; Fax: 44 (0)20-8392-9876
E-mail: Bushwd@aol.com
Free postage in the UK. Europe: Air mail at cost.
Please try your bookstore first.

Contents

Introduction

It's amazing what is out there; many people with children's automobiles simply do not know what they are worth or who to turn to for advice! Many fantastic finds came out of the woodwork for this book, I'm pleased to say. Hopefully with this publication, many more will be unearthed for the collector to see.

One of the most important parts of this book is the Tri-ang toy pedal car supplement. These pages were chosen from the many catalogues I acquired from an ex-Tri-ang technical director who had the foresight to keep most of the catalogues from his time with the company. By chance this particular person had come across my first book and showed a keen interest in how I acquired the ex- Prince William Bentley electric car that was produced especially for His Royal Highness during the 1980s.

The technical director had been employed by Tri-ang, and we arranged a meeting during the winter of 1997. He and his wife kindly brought all the catalogues down to me, and a price was agreed. I also purchased from him a

beautifully kept Tri-ang Rolls-Royce, based on the Silver Cloud III convertible, in its original box with three spare flying lady (the spirit of ecstasy) mascots for these cars. We had a wonderful day with him reminiscing about working for the Tri-ang company. The fruits of all this are included in the supplement.

Included here are official studio photos (shot in black & white) commissioned by Tri-ang to be used as reference in their own in-house archive folder. The archive also contains many examples of their other toys, including various wheel goods, bikes, trikes, prams, tin plate large cranes, trucks, cars, jeeps and tractors, as well as examples of rocking horses, swings, toy forts, dolls houses, and more. These toys all date from the 1960s period of manufacture, and this section makes for fascinating reading. The pedal car archive photos are included in this supplement.

I also have a few Mobo toy catalogues and wish to include these in a future volume. If you have any available for sale or that may be reproduced with compliments within the next volume to the supplier of material, I would be grateful.

With the great interest in the Internet, many collectors are up-loading websites on pedal cars and toy collecting in general. We too have our own website, and you can log on to the following:

http://www.Members.tripod.co.uk/G.G. Weiner
http://www.carsofbrighton.co.uk
http://www.eurosurf.com/cars
http://www.antiquesportfolio.com/pedalcars
http://www.brmmbrmm.com/pedalcars

I wish you great success with this hobby! Don't forget there is a club for you. Contact:

The Pedal Car Collectors' Club (P.C.C.C.)
4-4a Chapel Terrace Mews
Kemp Town
Brighton
East Sussex BN2 1HU
England
Phone/Fax: 0 1273 60 1960
www.brmmbrmm.com/pedalcars

With regard to the price valuations given, these are based upon market values and auction prices, and the values placed upon collectors', dealers', and traders' cars in present or past situations. The prices quoted are as accurate and up-to-date as possible based on the sources that provided them, bearing in mind the condition, rarity, and other prevailing factors of demand and supply, along with the market situation at the time of writing. However, the author and the contributors to this book take no responsibility for any action taken in buying, selling, or trading of cars by anyone consulting this book; it is intended only to be a guide.

Parade of Children's Automobiles

Austin Seven Chummy van of 1929 alongside a child's half-scale version. Wood panels were built onto a steel box section chassis with replicated radiator and working spring suspension, wire wheels, and rubber treaded pneumatic tires Powered by a Sinclair C5 electric motor with a 12-volt battery at the rear of the vehicle. *Courtesy of Keith Hawkins*, U.K.

1896 Lutzmann 4 hp Victoria, scratch built in half size with all controls replicated from the original car. It is powered by a single-cylinder 4-stroke petrol engine. Wooden body, curved dash with fancy wrought ironwork, steel sprung frame, wire wheels, solid rubber tires, pressed steel mudguards. All detailing as per original specification. Value undeclared. *Courtesy of Keith Hawkins*, U.K.

Extremely desirable Lines Bros. Ltd Rolls-Royce style pedal car, circa 1928. All-wooden construction with opening driver's door, pressed light gauge steel mudguards and running boards, large balloon steel wheels with rubber tire inserts, front sprung bumper steel steering wheel, wood seat on wood square frame (not a chassis at this period). Features louvered hood and impressive Rolls Royce style radiator grill with motor meter mascot, oil and petrol cans, special miniature tax disc, dummy headlamps and sidelights. Good, honest, untouched example. *Courtesy of Michael Keating* B.E., M.I.E.I., U.K. Value undeclared, estimated at £2,500-£3,000; $3,600-$4,350

MG 'Y' Type or Midget by (possibly) Lines Bros. Ltd., during the late 1920s to early '30s. Light gauge pressed steel body mounted on a wooden frame and back board to take the wooden seat, single opening door. All metal balloon wheels and steering wheel, painted radiator, MG octagon logo and number plate, and fitted aeroscreen (possibly not contemporary to this car). *Bob Wilnkler, U.K.* £950-£1,250; $1,400-$1,800

Bugatti Biplace en Tandem, an authentic 1920s Eureka pedal car for two children. This tourer is fully restored. The pressed heavy gauge steel body was prepared and repainted light blue. Other improvements include re-upholstered seats; re-chromed radiator, grab handle, and wheel hub caps; wooden dashboards; refurbished steel chassis; and replaced runningboards. *Courtesy of Auto Suggestion, U.K.* Value (restored) £2,500-£3,000; $3,600-$4,350

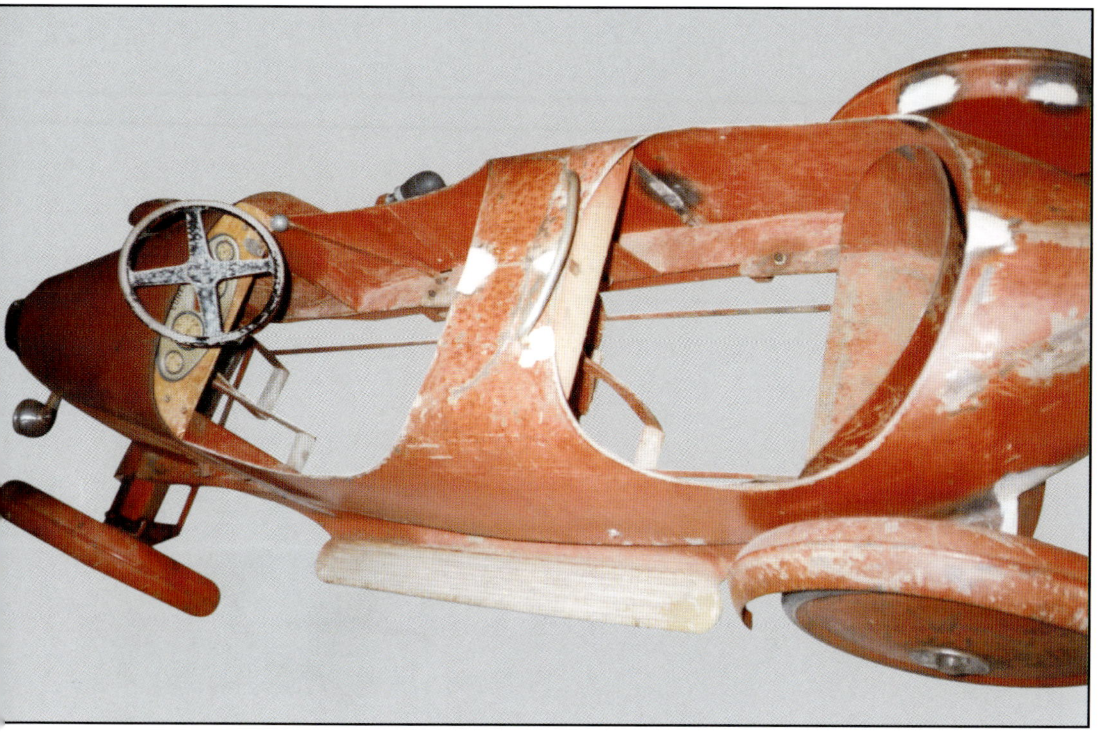

A similar Eureka Biplace en Tandem Bugatti in good original condition, having headlamps factory fitted to the radiator. Note authentic dashboard and handbrake, and original bulb horn to right side of car. *Courtesy of Gerard Breard, France.* £1,500-£2,000; $2,200-$2,900

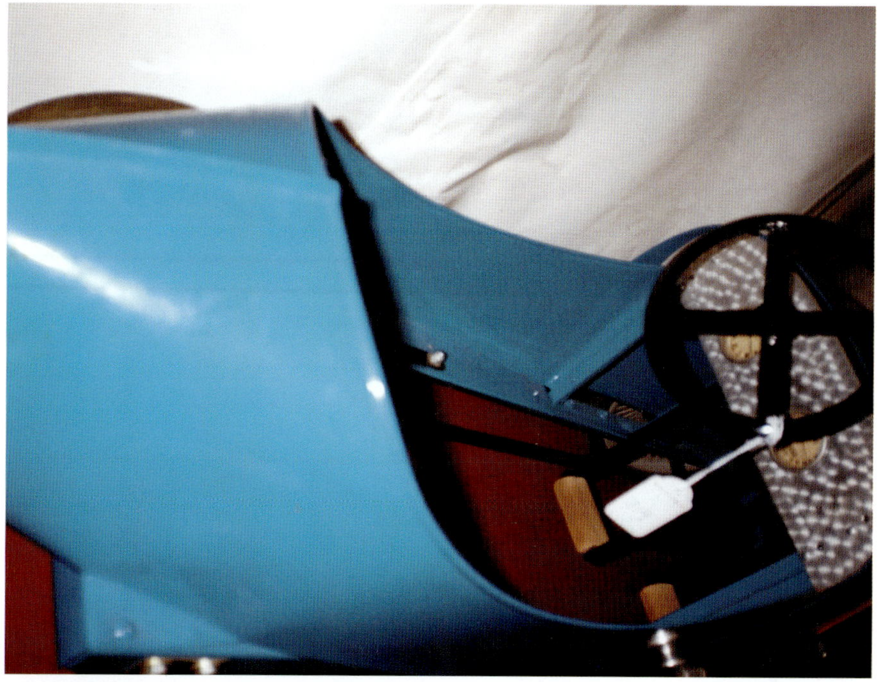

Top left and bottom left: Eureka Bugatti, circa 1928. All-steel body on metal frame, chromed radiator grill with enamel Bugatti badge, machine-turned alloy dash with dummy dials, non-original wheels and horn. Recently restored. *Courtesy Brooks Auctions, London.* £950-£1,250; $1,400-$1,800

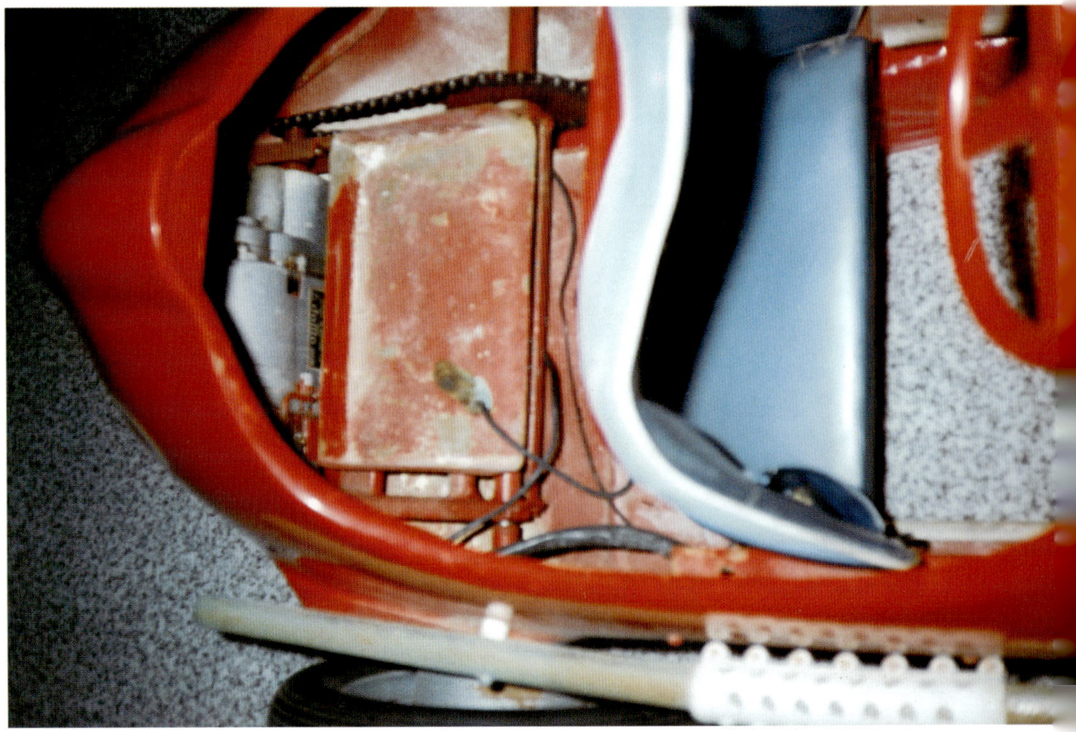

Amilcar C6 by Cercle Pegase Mailcar, circa 1920s. Child's electric replica with 12-volt battery operation to motor with chain drive. Heavy gauge steel louvered body on steel chassis frame, wire wheels with rubber treaded pneumatic tires, chrome dummy radiator and filling cap with rear petrol cap, steel steering wheel, plain dashboard. Features outside exhaust pipe, leather hood straps and drum brakes. Fully restored. *Owner M. Marteling, France.* Value undeclared.

French racing car (similar to Tri-ang Brooklands racer of the 1920s). The makers' name on radiator is 'Origine Jocker' with a jokers logo motive/emblem transfer. Light gauge pressed steel body, cut louvered vents, steel balloon wheels with inset rubber tires, steel steering wheel. Features outside working handbrake and radiator with dummy pressure cap. *Chris Holford collection*, U.K.. £550-£750; $800-$1,100

Jaguar 'D' Type Le Mans racing car of the 1950s, produced in moulded fiberglass on tubular steel chassis frame. Villiers four stroke petrol engine powering the rear wheels, automatic clutch, pull starter. Fitted with electric powered headlamps. Various team colours including British Racing Green, Ecurie Ecosse, Briggs Cunningham etc. Produced during the 1980s by Children's Classic Cars (U.K.). *Prop. A. Norton*. Values £1,500-£2,000; $2200-$2900

DELAHAYE-échelle 6/10
Full specification-along with price-in French.

CARROSSERIE

Fibre de verre/résine polyester
Peinture deux tons sur demande
Pare-brise rabattable
Pédalier réglable

CHASSIS

Structure tubulaire mécano-soudée

INTERIEUR

Sellerie assortie à la peinture deux tons
Tableau de bord en noyer

EQUIPEMENT ELECTRIQUE

Batterie 12 Volts
Phares / feux arrières
Klaxon
Démarrage électrique à clé
Coupe circuit extérieur

MOTEUR

Cylindrée : 319 cm3
Alésage x course : 76,2 x 69,8 mm
Puissance max. : 6,5 Kw (8,5 cv)

TRANSMISSION

Embrayage centrifuge
5 vitesses avant dont 4e et 5e condamnables
1 marche arrière

SUSPENSION

Essieu avant oscillant
Essieu arrière suspendu

ROUES

400-8 jantes acier

FREINS

Disque sur essieu arrière et frein à main de série indépendant

DIMENSIONS

Longueur : 2370 mm - largeur : 870 mm - hauteur 700 mm
Voie avant : 755 mm - voie arrière : 750 mm - empattement : 1520 mm

Automobiles Chambert se réserve le droit de modifier les caractéristiques techniques de ses modèles sans préavis.

Order cost in the region of F.50,000

Other views of this superb quarter scale version. Note one view alongside actual full size car. *Courtesy of Max Chambert, France*

Foreground: Panhard et Levassor, code EK10, £3,500-£4,000; $5,000-$5,800. Above background: Modele Super-Junior 35, code EK style Delahaye, £2,500-£3,000; $3,600-$4,350. Left background: Pedal speed boat, code EK21, £4,500-£5,000; $6,500-$7,200. Each dates in the 1920s-early 1930s.

Left: Modele Course 28 Grand Prix Code EK10, circa 1929. Right:1950 style Bugatti, £3,500-£4,000; $5,000-$5,800. Background: Modele Super-Junior 35, code EK19 style Delahaye, £2,500-£3,000; $3,600-$4,300.

Modele Biplace en Tandem 38, code EK22 style Delahaye, £5,500-£6,000; $7,900-$8,700.

Display of pedal cars by the Eureka Co., France at the Centre International de la Automobile, Pantin, Paris, France (a one-off private on-loan exhibition). Values were not declared with the descriptions and photos supplied by the above museum (to whom I am most grateful), and the ones that appear here are based upon the original factory fresh or professionally restored condition these pedal cars represent (inferior condition to this would of course be reflected in the value).

Foreground: Modele Biplace en Tandem 38, code EK32 style Delage deluxe, £9,000-£10,000. Background: Bugatti 28 Grand Prix Racer (based on type 35 GP same as photo 2), £3,500-£4,000; $5,000-$5,800

An automotive miniature mascot/hood ornament depicting a winged goddess holding a trophy laurel wreath for a (American?) pedal car of the inter-war period. Produced in electroplated nickel silver.

Left: Gendron 1938 Skippy Sharknose, U.S.A. All pressed steel monocoque body on steel chassis-frame with chromed accessories, replaced perspex windshield and side-wind deflectors. Vinyl upholstered seat. Finished in factory burgundy with white paintwork to fenders and chrome dummy fender lights. Chrome sprung front bumper, balloon wheels, chrome hubcaps with solid rubber inserted tires. Fully restored. £2,500-£3,000; $3,600-$4,300. *Courtesy C.A.R.S. of Brighton, U.K.*
Right: American National/Steelcraft Fire Truck of the early 1920s with pressed steel body, separate fender-running boards, original chrome grill, dummy headlights, front and rear chrome sprung steel bumpers, chrome bell and hood ornament. Steel body colour painted balloon wheels with rubber inserted tires. Replaced wooden ladder on hooks on windshield and rear grab handles all in light steel. Original steel steering wheel. Fully restored, repainted, new decals and coachlinings. £3,500-£4,000; $5,000-$5,800. *Courtesy C.A.R.S. of Brighton, U.K.*

Citröen 'Citrönette' child's replica used for promotions, publicity, and sales drives during the 1920s and '30s. Fully restored working replica with heavy gauge steel body, opening hood with 12-volt battery powered electric motor, chrome wheels with pneumatic rubber treaded tires, full leather upholstered seats, and authentic working dashboard. *Courtesy of Threeways Garage, U.K.* Value undeclared.

Austin Chummy by The Model Car Company, Louth, Lincs., U.K. Children's electric powered fiberglass body on steel box section chassis. £950-£1,250; $1,400-$1,800.

1931 Alfa Romeo 8C 2300 M. Scratch built child's replica of this famous marque in pressed sheet steel with aluminium floorpan, built on a steel box section chassis. Honda powered petrol engine under the opening louvered hood. Wire wheels with rubber-treaded pneumatic tires. Features authentic radiator with working calometer, enamel Alfa Romeo Milano hood badge, upholstered leather seat, working lamps and tail lights, spare wheel, leather retaining straps. Fully operational dials on turned alloy dashboard. *Courtesy of A.Newton, U.K.* Value undeclared.

The Cotswold Motor Museum: A marvellous collection of antique pedal cars, pedal aeroplanes, and other toys, including the famous T.V. personality 'Brum' the cheeky pedal car! Well worth a visit when you are in the U.K. to see a good selection of Lines Brothers products.

Atco trainer, official government project to offer driving lessons to youngsters in the U.K. during the inter-war years. Large heavy gauge steel construction with petrol engine, proper steering gear change, brakes etc. Featuring vinyl upholstered seat, outside handbrake. This example is on display at The Heritage Motor Museum, Gaydon, near Birmingham, U.K.

Large heavy gauge steel pedalcar by Ace, Birmingham, U.K., circa 1948-50. American style 'Police Chief' car with steel hollow cast balloon wheels with rubber inserts (the hubs are missing), bumpers with overiders, dummy grill, headlamps, steel steering wheel, wire windscreen, and chrome rear grab handles to boot (lid missing). Features working handbrake and unique steering system with rear suspension. *Courtesy of Jullian Moss Tyrell, U.K.* £500-£650; $700-$950.

Austin J40 types, showing two late versions in "barn or garage find" condition with rusted bodies and parts missing. For this relic condition expect to pay between £100 to £250; $150-$400. *Author's collection.*

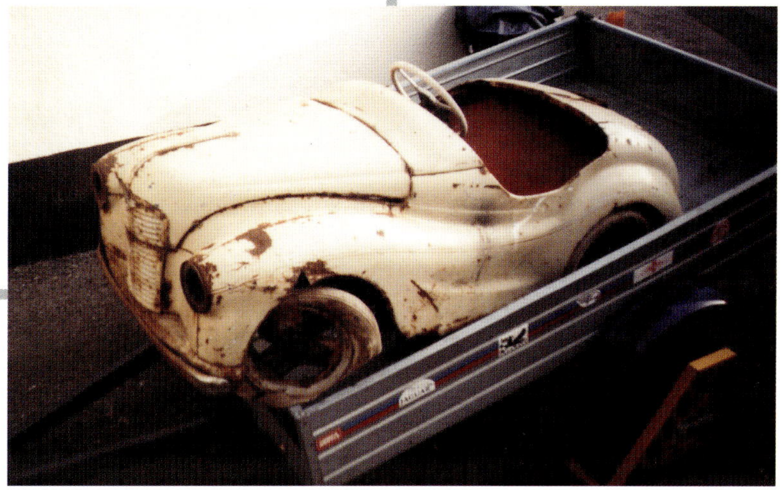

The restored early J40 with the 'A' hood ornament in place. This highly desirable state of restoration is worth investing in, and fetches £750-£1,250; $1,100-$1,800, depending on whether it's a private acquisition, or from a dealer or auction house.

Artwork for the Austin J40 & Pathfinder official factory sales brochure, produced by Robert Johnston during the 1950s.

Austin J40 fully restored early version with flying 'A' hood ornament.
Driver: Chico Golt of Brighton (U.K.)

J40: A Brief History

Bargoed, South Wales, in the late forties was an area dependant on mining, and the loss of a job in this field meant almost invariably joining the dole queue. With this sad fact in mind, Sir Leonard Lord conceived the idea, in 1948, of creating a factory which would employ men disabled by pneumoconiosis, and so forced to leave the mines. Doing light assembly work they would produce toy motor cars.

The idea was realised in 1949, and the official opening of the Austin Junior Car Factory was 5th July 1949. There were 55 men in its initial employ and the workforce gradually increased until it stood at just over 500 men in 1965.

They started production by making one model, the Pathfinder Special. It was based on the Austin 750 01-IC racing car. Its prototype was tried and tested, whereas the prototype for the forebears of the J40 was not satisfactory, so it was the Pathfinder which was produced first.

The J40 prototype, called "Joy I," was designed with the Austin 10 in mind, (with separate headlamps), but as the Austin 10 was going out of fashion, it was thought better to design and produce a more modern looking roadster. So Joy 4 was based on the A40 Devon. Production of this last prototype, to be called the J40 roadster, was ready to go ahead, so after one year, the Pathfinder was dropped. No record remains of the number of Pathfinders made.

The J40 production was in the later years rather seasonal in its demands, peak being the weeks before Christmas. The method was similar to that of a full size car. Individual body panels were pressed out and then welded together to form the body shell. The metal used was thought to be off-cuts from Austins built at Longbridge, which made them very solid. This in turn was degreased, ready for painting.

The paint colours used, although all thought to be Austin colours (or from the BMC/BL group), were constantly changing. The painted shell then passed along an assembly line to be fitted with its pedals, running and steering gear, wheels and tires, spark plugs, working hand brake, upholstered seats, bumpers and number plates, grille, badge and other chrome fittings, and of course its working headlights and horn.

Production ceased in 1971, 32,098 cars having been built.

Reproduced with kind permission from the Austin J40 Pedal Car Club circular.

Austin J40
Two careful owners from new. Never raced or rallied. Reason for sale: now old enough to go out with girls, so I need the money!
£2850 ovno
SUPERCAR CLASSICS November 1989

What used to be! An Austin J40 being advertised for sale during the boom years of the late 1980s... Remember, prices can go down as well as up!

Ferrari Indy racing car, superb one-off handbuilt half scale version in all respects to full size car, Honda petrol powered 72cc engine as raced by Nicki Lauda/Clay Regazoni. Value £6,500; $9,400. Anon.

Austin J40 converted into a 1950s Humber look-a-like during a complete restoration with two-tone re-paint. Seen on the Humber Owners' Club stand at the N.E.C. Classic Car Show, U.K. in 1997. Value undeclared.

Left and below: Brighton Southdown Co. open top bus. Scratch built wood body formed onto a steel box section chassis, powered by an electric motor taken from an invalid carriage. Value undeclared. *Ben Streeter*, U.K.

The Harrington Coach Co. (coach builders) Hove, East Sussex. This rare example of a 1960s scale model of the full size coach supplied to the Southdown Coach Co. was built to promote the company and used along Brighton and Hove seafronts. Powered by a Villiers petrol engine, heavy gauge metal body on ash wood frame, authentic colours, decals and chrome trim, working lights and horn, pneumatic rubber tires and chrome wheel hubs. The driver (single seater) sits in the middle of this example. *Courtesy of the Chris Booth collection, Rolvenden, Kent, U.K.* Value undeclared.

More views of Aramis de Macedo Secundino in his museum with a line of restored J40s, models and bikes.

Left: Pedal car lover, restorer, and well-known personality Aramis de Macedo Secundino as man and boy, pictured with an extremely rare and unusual pedal car by Brinquedos Gravina Toy Co. of Brazil, c.1939. Pick-up truck in pressed heavy gauge steel with wooden storage compartment at rear, balloon steel wheels, light pressed steel mudguards and bumpers, steel steering wheel on steel chassis frame, complete with re-upholstered vinyl seat. Restored in 1998 by Aramis de Macedo Secundino. Value undeclared.

Citroën 2CV child's pedal car. Authentic replica in fiberglass with fine detailing, featuring foldback canvas roof, upholstered seats, working lights etc. Other specifications not known, as the car was photographed outside display area at the Paris Retromobile in February 1997.

Tri-ang Zephyr model Thirty-B, early 1950s production in pressed steel, balloon wheels with chrome hubcaps, steel arrow logos, tin windscreen, steel steering wheel, re-upholstered vinyl seat, dummy grill and headlamps (horn a later addition). *Aramis de Macedo Secundino*, Brazil. £250-£350; $400-$500.

"The French Collection". A wide range of various French produced pedal cars from the 1920s to 1950s exhibited on Dexion shelving at The Paris Retromobile, 1997, by André Colas, Paris, France.

Lines Bros.Ltd.replicas, faithfully reproduced in light gauge pressed steel with wooden frame, opening left-hand door, chrome wire wheels with rubber treaded pneumatic tires, steel mudguards, authentic radiators with battery operated chrome headlamps, horn. Fine body detailing with perspex windscreen, steel Brooklands style steering wheel, authentic dashboard, vinyl upholstery, number plates and working handbrake. C. & A. Demler, Germany. £1,000-£1,500. Models shown are (above) Daimler Sports roadster, (below) Daimler, Bentley, MG sportscars.

Superb MG K3 child's replica powered by a Briggs & Stratton engine. Steel body panels with wooden rear, vinyl covered wood bench seat, built on a steel box section chassis frame, authentic wire wheels with rubber treaded pneumatic tires, working headlamps and horn. Features MG wire mesh grill, chrome radiator surround, outside exhaust, working gears and handbrake, and disc brakes. A superb one-off scratchbuilt replica by J. Hubbard, U.K. Value undeclared.

MG

Ford Model T. Superb scratchbuilt child's replica powered by a Briggs & Stratton engine. Photos showing the wooden body with wire wheels and rubber treaded pneumatic tires, authentic dashboard, hood, and grill. Superb gearbox detailing mounted on a steel box section chassis frame. A one-off scratchbuilt replica by J. Hubbard, U.K. Value undeclared.

Tom Thumb (modern) replicas, Holland. Each of these Bugatti Type 35s sports a pressed steel body with louvered bonnet, wire wheels with rubber treaded pneumatic tyres, built onto a steel chassis, wood steering wheel, authentic dashboard, working outside handbrake, fitted with twin aeroscreens. Featuring working lights and vinyl upholstered seats. Prices £1,250-£1,500; $1,800-$2,100.

Panhard et Levassor – *Courtesy of L'automobiliste Issue No. 14.*

Authentic handbuilt (modern) French Eureka style replicas of The Panhard et Levassor and Bugatti Type 35 in pedal car form. Constructed of pressed steel bodywork on a steel chassis frame with chromed pressed steel wheels with rubber treaded pneumatic tires and authentic working suspension. Has fly-off handbrake with gear change mechanism. Option of working headlamps and horn. Elantec models, U.K. £3,500-£4,000; $5,000-$5,800.

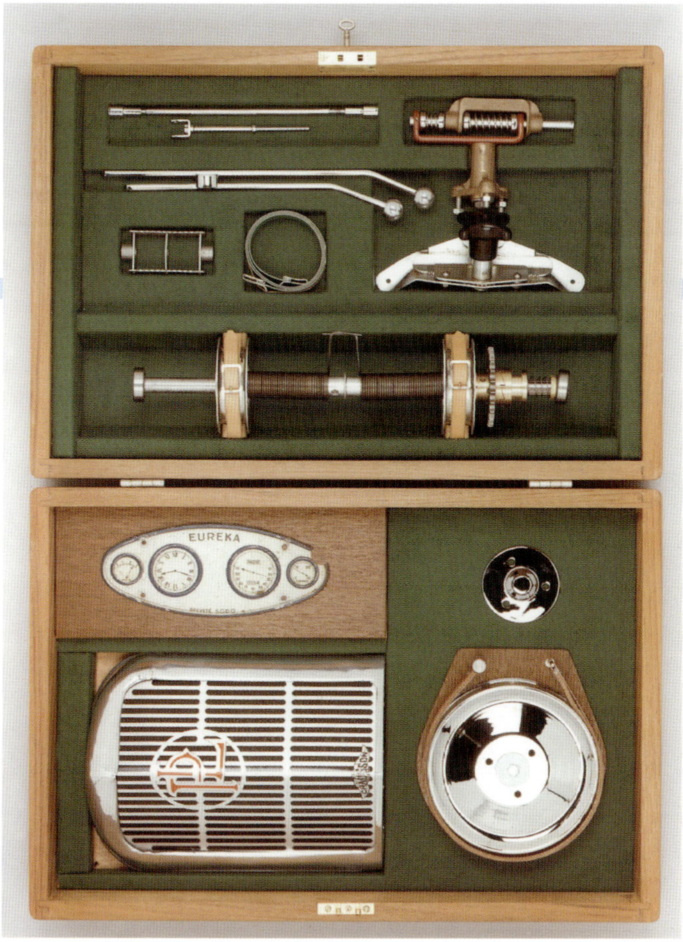

Eureka style Panhard et Levassor by Elantec Replicas, U.K. A superb replica of this 1920s car produced in pedal car form showing remarkable detail in the heavy gauge pressed steel body with louvers, chrome radiator, bumpers and wheel hubs, rubber treaded pneumatic tires with cycle wings. Features vinyl upholstered seat, working headlamps and horn, authentic gear and handbrake sticks (to inside of car) Complete with spare radiator shell, dashboard, and accessories in fitted wooden case. Made to order from £4,000; $5,800.

Bugatti Type 35, petrol driven with possible 50cc motor. All heavy gauge pressed steel body with removable hood for battery compartment engine shown at rear under lift-off vinyl seat. Featuring aeroscreen, turned alloy dashboard with working dials, outside handbrake, alloy wheels with rubber treaded pneumatic tires. Maker unknown (probably French origin) value £2,000-£2,500; $2,900-$3,600. *Courtesy Brooks Auctions, London.*

Bugatti Type 52 (based on the Type 35 Grand Prix car of the 1920s) faithful replica 12 volt electric powered child's car by Crosswaite & Gardiner, Ltd., U.K. *Top left*: *Courtesy of Sotheby's*, London. *Bottom left*: Driver Sam Womersley. *Below*: *Author's collection.*

Giordani of Italy Formula 1 style racing car of the mid-1950s. Large heavy gauge, steel monocoque body shell, wire mesh steel grill with inset dummy lights, wire spoked wheels with ear spinners, dummy exhaust. Steering wheel and perspex windscreen missing. *Courtesy J.M. Venning, U.K.* £450-£500; $650-$750.

Exhibits from the Glasgow Museum of Transport which include pedal cars and bubblecars, well worth a visit if ever you travel to Scotland.

Tri-ang Ford style 1960s production in light gauge pressed steel, all steel parts including steering wheel, balloon wheels, rubber tire inserts, dummy grill, chromed front bumper. Features dummy gear shift column change and (non original) driver' s spotlight. Restored condition. *I.Tapley collection, France.* £350-£400

M.G. (France) Honda racing car (similar to Tri-ang Vanwall of same period) in heavy gauge pressed steel, alloy wire wheels with solid rubber treaded tires, steel steering wheel, moulded plastic seat, original decals of the period, 1960s production. *I.Tapley collection, France.* £150-£200.

Tri-ang pick-up and dump trucks, showing the various body styles adapted from the Ford style Zephyr of the 1950s to early 1960s period of manufacture. This period featured all pressed heavy gauge steel construction with no plastic parts. Dummy radiator grill, front bumper with over-riders. The car on the left shows the loading area at rear with wooden inner frame. At right is the dump-truck version with simple lever to raise or lower and swivel the shovel. All types are quite hard to find now. *Top and bottom left, courtesy of Brian T. Krieger, B.C. Canada. Below, right, Author's collection*. Values between £100; $150 for relic condition up to £350; $500 fully restored.

M.G. (France) 1960s replica Citröen DS, in ready-to-restore condition, the pressed steel body rubbed down and prepared for painting. Original parts present including working headlamps. *I. Tapley collection, France.* 'As is,' £100-£150; $150-$200, when restored approx. £300-£350; $450-$500.

Left to right: Le Mans Esso racing car, Le Mans Porsche, Mercedes 500SL, Mercedes SLK, Mercedes 500SL (small version), Jaguar XJ6, Ferrari F1 Grand Prix, Triumph TR4. All of the 1980s to 1990s, the TR4 produced in the mid 1960s by Tri-ang all the rest are Italian made by the Toys Toys Co. Milan.

A selection of toys and pedal cars old and new. Top photo (left to right): Tri-ang Moulton patent child's bike c.1968; Tri-ang Ford style 1950s; Ace, Birmingham U.K.-made American style police car, 1948; Mobo, 1960s; Leeway police patrol car 1960s; Giordani, Italy, racing car 1950s; Mobo U.S.Army jeep 1960s; Lines Bros.mono pedal plane 1920s; Mobo rocking horse 1960s. Top row on boxes as described above.

HAMILTON JEEP Manufactured by the Hamilton Steel Products Company located at 1845 West 74th, street in Chicago, Illinois USA. This jeep was made sometime between 1950 and 1966 and was lettered U.S. Air Force. It was manufactured and painted a rather dull military blue. Jim Crean did a total "ground up " restoration project and restored it to a four-star General's U.S. Army jeep. This jeep was never manufactured as pictured but was restored to the owner's liking! Value in the U.S.A. at $1,800.

The Red Baron flies again! This tri-wing pedal airplane was designed and developed by Jim Crean in his studio in Hamburg, New York, U.S.A. It was manufactured by Paul Ezra of Metal Fab Engineering of Winamac, Indiana, U.S.A. It was created by Jim Crean after the famous Fokker Dr1 Tri-wing. It was flown during World War I by the famous German flying ace Manfred Von Richthofen, who was the most feared by his aerial enemies for his speed and the use of the two machine guns he had mounted on the wings. You will notice that the pedal plane also has duel machine guns.

Baron Von Richthofen has 81 "kills" to his credit in his Fokker Dr1 aircraft. He was killed in aerial combat on Sunday April 21st, 1918, shot down by R.A.F. Captain A. Roy Brown over the Somme battlefield. Paul Ezra took this pedal airplane to the National Pedal Car Show held in Kalamazoo, Michigan, USA, in November of 1996 and won first place in the "Peoples Choice" award and first place in the Military Restored class.

The pedal aircraft is manufactured out of 18 gauge steel and is 52" long—with a wingspan of 34". It is 24" high its weight is 52 pounds. It was manufactured in the United States in a limited number and sold for $3,500, plus freight and shipping charges.

If you are interested, a special limited edition for the European market could be manufactured. Please contact: Jim Crean, 4876 Big Tree Road Hamburg, New York 14075 U.S.A. for further details.

Jaguar XKSS produced in small numbers by The Glassfibre Co. Birmingham, England during the 1950s. Powered by a Briggs & Stratton petrol engine (located in the opening trunk). Built on a steel box section chassis frame, perspex windscreen, working foot pedals, two-speed gearshift with reverse. Moulded plastic steering wheel alloy wheels with rubber treaded pneumatic tyres. *Collection of Tony Plowright, U.K.* £3,000-£3,500; $4,350-$5,100.

Pink Cadillac, typical 1950s "tails & fins" scratchbuilt fiberglass three-quarter size version, alloy wheels with pneumatic rubber treaded tires, vinyl upholstery. Features working radio, lights, indicators and horn, authentic dashboard, and a trunk that opens to reveal a small petrol engine. *Collection of Tony Plowright, U.K.* £1,250-£1,500; $1,800-$2,200.

Mid 1960s Tri-ang Rolls-Royce Silver Cloud III, convertible. Moulded plastic monocoque on metal support frame, chromed metal radiator grill, mascot, bumpers, horn vents, wheels and hubs, with metal steering wheel and perspex windscreen. Has working lights and horn from a small battery. Dummy plastic drop-down hood and plastic seat. *Norman Gibbs, U.K.* In its original cardboard carton, £350-£500; $500-$750.

Chitty Chitty Bang Bang fantasy pedal car by Pines of Italy, 1968-1970. Moulded plastic body with pull-out wing sections, running boards, metal steering wheel, spoked wheels, rubber tires, (some parts missing including perspex windscreen). Metal radiator, headlamps, outside exhaust and GEN 1 number all contemporary to the car. £500-£750; $700-$1,100 in this condition.

Lines Bros. Ltd. Rolls-Royce replica. Barker type bodywork in wood on metal chassis frame, opening driver's door, running boards, non-original petrol can, opening alloy windscreen frame, glass windscreen. Has authentic polished steel grill, working headlamps, horn, wire wheels, rubber treaded tires, mudguards. Steel steering wheel, drop-down dummy hood frame, leather upholstery. Overly restored, however a rare example. Anon. £1,250-£1,500; $1,800-$2,200.

Tri-ang "Motorway patrol" car, late 1960s-early '70s. Pressed light gauge metal, integrated polished grill, bumper, rear lights and bumper, steel balloon wheels with inset rubber tires, plastic steering wheel. Features flashing blue light on hood and horn/siren powered by a small battery under the hood. *Courtesy of Mark Josey, U.K.* £100-£150; $150-$200.

Triumph TR4/4A by Tri-ang Toys 1960s production in moulded plastic forming a monocoque over a metal frame, steel wire wheels, windscreen, front grill and bumpers, dummy headlamps and sidelights (replaced steering wheel, rear bumper missing). In fair condition. *Courtesy of RobSport, U.K.* £150-£250; $200-$350.

Half-scale motorized Triumph TR3 (no motor on this example) child's replica in fiberglass with non-working lights and horn. Authentic dashboard, vinyl upholstered seat. Triumph badges on opening hood and trunk. Proper wheels and rubber treaded tires. *Classic Cartrim, U.K.* Estimated value £950-£1,250; $1400-$1,800.

Ferrari F1 Grand Prix racing car, based on a 1950s racer. Produced with fiberglass moulded body on a steel chassis with moulded seat, wooden steering wheel, alloy wheels, rubber treaded pneumatic tires, and wire mesh grill with chrome trim. Powered by a Kawasaki 5-hp engine in rear, automatic clutch, single drive, two gears. Limited edition of 10 only, this being No. 8. *Courtesy of A. Walter*, U.K. Value £1,500-£2,000; $2,200-$2,900.

Late 1980s Ferrari F1 Junior Grand Prix half-scale model, accurate replica of Nigel Mansell's car (this one built especially for his son). Moulded fiberglass shell built on to a steel chassis frame, ignition switch electric starter, Briggs & Stratton air cooled 400cc 4-stroke engine. Features full seat harness, removable steering wheel, and disk brakes. U.K. built. *Courtesy of Keith Wareham, U.K.* Value undeclared.

Sila (France) V12 Bimbo racer, large moulded heavy gauge steel monocoque on a steel chassis frame with wire wheels, hub spinners and pneumatic rubber treaded tires. Perspex windscreen, plastic steering wheel, dummy egg crate grill with hood badges, some parts missing. Driven by 12 volt electric motor. *Collection of Jean Michel Mabileau, France*. Value in French Francs F22,000.

1950s Ferrari Testarossa racing car. A modern made child's version, produced in fiberglass on a steel chassis frame, driven by a petrol Tecumsi (U.S.A.) engine. Very fine detailing, including Borani style wire wheels, chrome side exhaust, working lights, authentic dashboard, perspex windscreen, and tonneau cover. Modern product. *Owner Nick Mathews, U.K.* £5,000; $7,250.

Tri-ang police type jeep and Porsche 911 type pedal cars in moulded plastic, with plastic wheels and tyres; the only metal is the frame that holds the body. Late 1970s to early 1980s. *Courtesy of Paul Wilcox, U.K.* £45-£65; $70-$100 each.

Little Tikes buggies dating from the 1980s (similar to present day). Entirely produced in moulded plastic without any metal parts. *Courtesy of Paul Wilcox, U.K.* £30; $40 to current prices.

A Magic Roundabout indeed! Comprised of different coloured Austin J40 fiberglass replicas, last located on Brighton beachfront, Sussex coast, U.K.

Modern moulded plastic pedal and electric six-volt powered replicas of many famous marques of cars and fantasy vehicles. *Italian made by the company Toys Toys, Milan.*

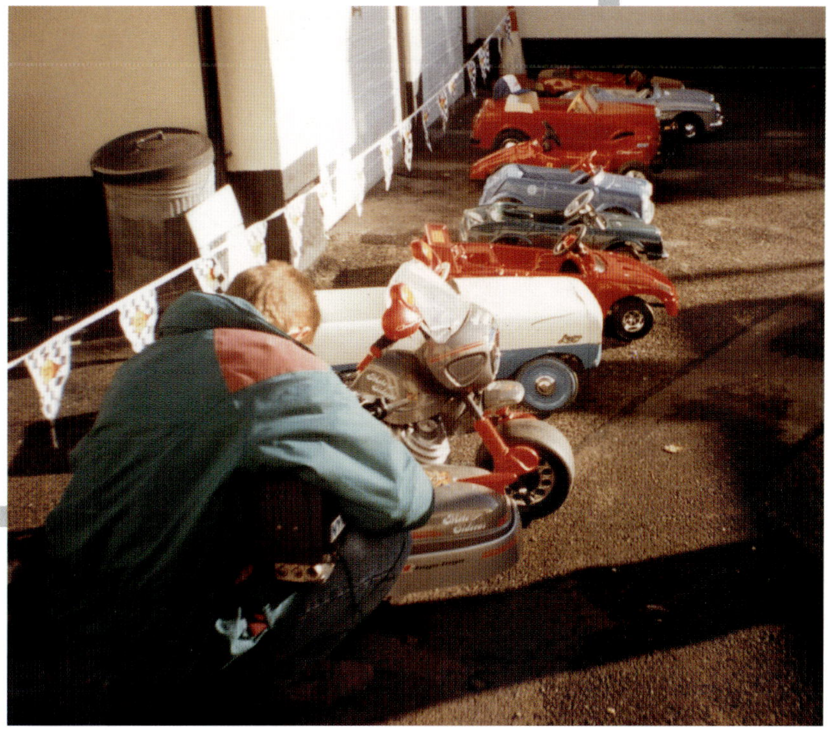

Filming at C.A.R.S. premises in Brighton, Sussex, U.K., for the television program *The Line* by Nynex/Cable & Wireless Co., on 25 November, 1996 for pre-Christmas show.

MG TF Series half-size replica in fiberglass monocoque construction. Twin seater with steel chassis, steel wheels with pneumatic rubber treaded tires, chrome hub caps, perspex fold down windscreen, authentic working dashboard with turnkey operation for 12-volt electric motor. Modern production by Sunny Day Car Co. U.K. *Children courtesy of the Day family.* Prices from £1,250; $1,800 depending on specification order.

MGBs by Abingdon Model Co., U.K. Authentic fiberglass body built on a steel square chassis frame, pedal powered, with solid rubber tires on disc wheels. Features vinyl seat, steel MG logo steering wheel, perspex windscreen, working lights and horn. Modern production in two basic colors of British racing green and signal red. £750 to order.

Mercedes 500 SL. Motorized children's car in moulded fiberglass on steel chassis frame with working gear change, leather upholstered seat, working lights and horn, proper wheels and tires, authentic detailing. No other specs provided. *Courtesy Brooks Auctions, London.*

Fiberglass child's scale Audi sports model. Two-seater with motorized petrol engine,vinyl upholstery, authentic working dashboard, foot pedals, and gear change. Built on a steel chassis frame with perspex windscreen, alloy wheels, and rubber treaded tires. Fine detailing. Further specs not known. Photograph taken at Earls Court, London, October 1997.

Jaguar XK120 half-scale replica. Fiberglass body, steel wheels with rubber treaded pneumatic tires, built on a steel box section chassis frame, authentic dashboard with turn key operation for Briggs & Stratton petrol engine. Features detailed working headlamps, tail lights, horn, perspex windscreen, alloy fittings. A superb replica photographed in France. Value not given, but as a one-off £5,000-£6,500; $7,250-$9,500.

MG Magnet ZB of the late 1950s. A faithful replica produced in moulded fiberglass, metal wheels with solid rubber tires, perspex windscreen, upholstered material seat. Features working headlamps. A self-build /scratchbuilt car by Peter Bonthrone, U.K. Value undeclared but instruction manual available from him by sending a payment of £5 to: *Children's Classic Pedal Car Construction, 35 Wantage Road, Wallingford, Oxfordshire, OX10 0LR, U.K.*

A Ferrari F1 Pedal Car (Tri-ang, 1970s)

The following six cars are moulded plastic with minimum of metal parts, produced from the 1970s (photos A & B) to the present day (photos C to F). These children's toys are quite common, the types dating from the 1970s can be acquired in the £75; $100 to £100; $150 range if in good condition with no parts missing. The modern pedal cars and electric powered cars are readily available from various outlets and no more than the current list price should be paid. *Courtesy of Paul Wilcox*, U.K.

B Ferrari F1 Pedal Car (Tri-ang, 1970s)

C Williams Renault F1 (Smoby, 1997)

E Porsche Baby Boxster (Toys Toys, 1997)

D Porsche 911 Carrera Turbo (Giordani, 1980s) 6v battery car

F Porsche Boxster (Toys Toys, 1998)

Postcard photograph from the Norsk Teknisk Museum of the same car.

Pedal cars of Norway are included in this book as no previous mention or reference to them has been made (to my knowledge) before. The author is greatly indebted to Mr. Ole Birger Gjevre of Strand, Fagernes, Norway for kindly supplying the information along with photos and his opinion of valuations in Krona. *With grateful thanks to The Norsk Teknisk Museum, Norway.*

Mr. Birger Gjevre alongside a 1912 Cadillac electric powered two-fifths scale representation made especially for Crown Prince Olav of Norway. This car has a real leather button back seat, folding hood, glass drop-down windscreen, opening doors, and fine detailing with working lights and steering just as in the full size car. Built by J. Lockwood & Co., Islington, London, U.K.

The U.K. Cadillac concessionaries F.S. Bennet commissioned the London coachbuilders J. Lockwood & Co. of Islington to construct a two-fifths scale electric powered replica of the full-size latest Cadillac to display on their stand at the 1912 Paris Motor Show. It caused great interest, attracting crowds around it. It was reserved for Queen Alexandra (wife of King Edward VII) and sent as a gift to her nephew Crown Prince Olav of Norway. In 1948 he presented it to the Norsk Teknisk Museum where it is still on permanent display. This car is one of only two produced. The other was commissioned by Prince Rama of Thailand and presented to his son Prince Chula Chakrabongse of Siam, which was on display (until 1998) in The London Toy & Model Museum.

'Tarzans Trabil, Lillestrom, Norway'. These pedal cars were unusual as they were produced in plywood over a metal frame with steel steering wheel, steel spoked wheels, and padded seat (probably vinyl). The ones in the photo probably have spring metal bumpers fitted. These were produced in large numbers during the 1950s.

'Grepa & Mjelva A/S, Orsta, Norway Go-Kart (No. 20) and cars as described previously. All models are of pressed steel, the early ones having steel-spoked wire wheels with solid rubber tires later changing to pneumatic. The firm existed from 1932 until the early 1970s.

The Christmas postcard showing Her Royal Highnesses Princess Ragnhild and Princess Astrid with a 1930s Grepa model No.1 pedal car.

'Sigurd W. Odegaard, Oslo, Norway' Jeeps marketed as 'Jeepy' and produced in pressed steel during the period 1946 to 1952. The first models had solid rubber tires, then pneumatics were introduced. The seats were adjustable to suit the size of the child. The snow plow and number plates are not contemporary to the models.

From the collection of Ole Birger Gjevre.

Grepa pedal car in pressed steel with steel balloon wheels and solid tires. A very simple car from around 1972, produced by 'Grepa & Mjelva A/S, Orsta, Norway.'

'Sigurd W. Odegaard, Oslo, Norway' Jeep in pressed steel with pneumatic rubber tyres.

Checker taxi cab made in present-day China, representing a Murray (U.S.A.) pedal car of the late 1940s to early 1950s. Retailed by Burns Novelty Toy Co., LLG, California. Retailing for about $300. *Courtesy Ole Birger Gjevre Collection.*

Grepa replica pedal car in pressed steel with toy car alongside produced by Lars Bekk and Jan Backe in 1990-91. Only sixteen of these pedal cars were produced, and the company name is stamped under the seat "B.B.Tabilverksted." They were retailed for N.Kr. 2,500. The 1/4 scale toy car was a marketing promotional product and stamped "B. B. Pedalcarworkshop."

Tvistegubben, produced in the 1960s by the firm 'Tvistegubben, Moller Lien, Kongshan, Tromoy, Norway.' Moulded plastic built on to a square chassis metal frame with wire spoked wheels, solid rubber tires, and metal steering wheels. Production finished in the early 1990s.

Touchwood faithful replica of an M.G. 'T.D.' (denoted by separate headlamps) in all-fiberglass monocoque construction with chrome plated front and rear bumpers, grill and headlamp surrounds, aluminum windscreen in fixed position. Has authentic dashboard with dummy dials, metal steering wheel and on/off button for headlamps (the battery under the non-opening bonnet). Metal solid wheels with chrome hubcaps with spare wheel to boot. Upholstered in leather-look vinyl. Modern made replica. T.W.M., Norwich, Norfolk, UK. £650-£700; $950-$1100.

Steelcraft Toy Co. American produced in the 1930s, popular replica pedal fire truck. Fine detail, having ladder, bell, working headlamps. Fully restored, coachlined, and new logos. Very desirable among American toy collectors. £950-£1,250; $1,400-$1,900.

Tri-ang Toys 1960s produced veteran car (loosely based on the French De Dion Bouton of the early 1900s). Heavy gauge pressed steel panels on tubular frame.

1953 Cadillac Eldorado convertible of the early 1950s. U.S. manufactured, possibly by Gendron Toys. Pressed steel body; steel wheels; chrome hubcaps, rubber-treaded tires, body trim, grill, and bumpers; authentic vinyl upholstery, perspex windshield, working headlamps and horn. Value undeclared by owner (who has fully restored this) but estimated at £1,500-£2,000; $2,200-$3,000.

Tri-Ang Toys Vanwall Grand Prix racing car of the early 1960s (produced into the late 1960s) in pressed steel with molded plastic seat and steering wheel, chromed steel wire wheels, exhaust pipe, authentic sprung suspension, solid rubber tires, perspex windscreen. Both this car and the Cadillac on the facing page were lovingly restored by their Brazillian owner Aramis de Macedo Secundino. Value undeclared but estimated at £350-£500;$500-$750.

Jeep tow truck, manufactured by the Hamilton Mfg. Co., U.S.A., under licence from Willys, U.S.A. All-steel construction, fully restored and repainted for advertising and promotional purposes (originally had logos of 'Official ATA' service car on the sides). Photos showing original pre-restoration as a 'relic', value £150-£200 ($200-300); and as fully restored and repainted to owner's liking £1,000-£1,500 ($1,500-$2,300).

1948 Pontiac convertible roadster, manufactured by the Murray Mfg. Co., Cleveland, Ohio, U.S.A. Produced throughout the 1950s. All pressed steel construction with chrome fittings. Fully restored. £1,000-£1,250; $1,500-$1,900.

1949 Buick Torpedo convertible roadster, manufactured by the Murray Mfg. Co., Cleveland, Ohio, U.S.A. Produced throughout the 1950s. All pressed steel construction with chrome fittings and bulb horn. Fully restored. £2,000-£2,500; $3,000-$3,700.

1951 Taxi Cab (retailed as 'dipside champion'), manufactured by the Murray Mfg. Co., Cleveland, Ohio, U.S.A. Produced throughout the 1950s to early 1960s. All pressed steel construction with chrome fittings and bulb horn. Fully restored with replacement decals. £1,000-£1,500; $1,500-$2,200.

1953 'Hot Rod', manufactured by the Garton Toy Co., Sheboygan, Wisconsin, U.S.A. All pressed steel construction. Chain-drive pedal operation with non-original chrome greyhound hood ornament. Fully restored and repainted (originally all yellow). £950-£1,250; $1,400-$1,900.

Gendron 1938 Skippy Sharknose, U.S.A. All pressed steel monocoque body on steel chassis-frame with chromed accessories, replaced perspex windshield and side-wind deflectors. Vinyl upholstered seat. Finished in factory burgundy with white paintwork to fenders and chrome dummy fender lights. Chrome sprung front bumper, balloon wheels, chrome hubcaps with solid rubber inserted tires. Fully restored. £2,500-£3,000; $3,600-$4,300.

Tractor produced by Tri-ang in the mid-1960s. Plastic body, pedal-powered rear wheels, molded deep rubber treaded tires. Metal seat, mudguards, and body frame. Features dummy gearshift and radiator grill. *Courtesy of Derek Eaton, U.K.* £150-£200; $225-$300.

Stutz open roadster by Steelcraft, c. 1929. Heavy gauge steel on steel box section chassis frame, runningboards, and sprung front and rear bumpers. Chrome radiator, dummy pressure cap motormeter, enamel "Stutz" badge, working headlights, sidelamp, and horn. Chrome windshield with perspex screen, dummy gauges on dashboard, vinyl upholstery, and fold-down top. Beautifully restored with painted coachlines. *Courtesy of the Warwick Henderson Gallery, New Zealand.* £1,500-£2,000; $2,200-$3,000.

Open sports tourer by Lines Brothers, Ltd., c. 1937. Medium gauge pressed steel on a box frame steel chassis. Single-opening door with chrome handle, chrome side ornaments, windscreen surround with perspex windshield, steel and chrome steering wheel, detailed radiator, front bumper (only). Steel mudguards with balloon wheels, solid rubber tires, and chrome hubcaps. Vinyl upholstery with dummy canvas folded top. Beautifully restored. *Courtesy of the Warwick Henderson Gallery, New Zealand.* £1,500-£2,000; $2,200-$3,000.

1933-34 Skippy Fire Chief in heavy gauge pressed steel. Authentic chrome radiator with enamel badge, chrome bell, dummy headlights, and working siren. Heavy steel fenders with balloon wheels, solid rubber tires, chrome hubcaps, and steel steering wheel. Featuring authentic dashboard, reupholstered vinyl seat, and reworked logos. *A beautifully restored prizewinning exhibit owned by Veronica Doster, U.S.A.* £5,500-£6,000; $8,200-$9000.

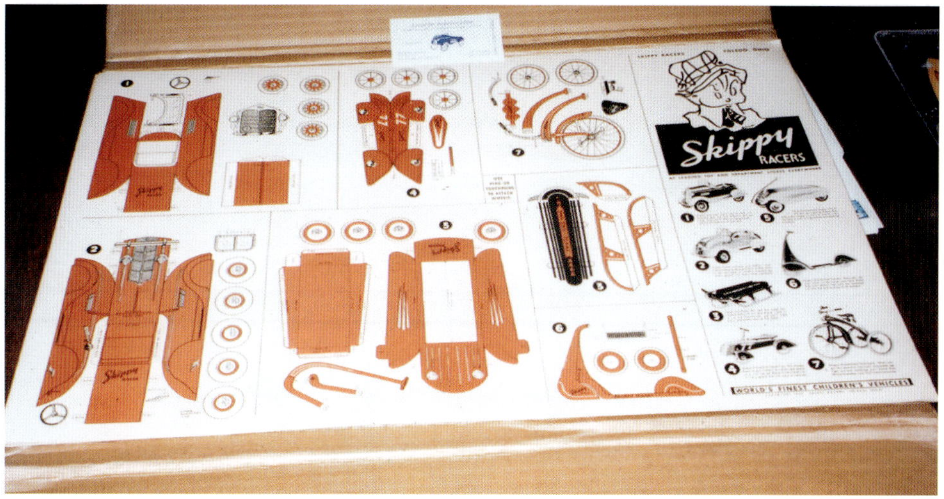

Original Skippy 1935 factory wheel goods advertisement sheet. *Courtesy of Matthew Vaznain, U.S.A.*

Station wagon, produced by Murray, c. 1948-1950. Medium gauge pressed steel incorporating rear storage compartment with grab handles. "Sad face" dummy grill, steel balloon wheels, solid rubber inserted tires. Restored and repainted in one color. £200-£250. Alongside this is a modern production miniature version, £35-£40; $50-60. *Owner, Ned Darr, U.S.A.*

Left: Skippy Racer Wagon, c. 1935. Heavy gauge steel, chrome pull-along grab handle. Steel balloon wheels with solid rubber tires, chrome hubcaps, dummy painted radiator, and chrome headlights. Expertly restored example. *Courtesy of Matthew Vaznain, U.S.A.* £750-£1,000; $1,100-$1,500.

1956 Murray *Dragnet* car, based on the popular T.V. show of the time starring Jack Webb. Medium gauge one-piece pressed steel body, dummy radiator with front bumper. Steel balloon wheels with whiteband solid rubber tires and chrome hubcaps. Has "Dragnet" decals on sides of body. Fully restored. *Courtesy of Richard and Ethel Seiverling, U.S.A.* £1,250-£2,000; $1,800-$3,000.

1960s Mobo police car in one-piece light gauge pressed steel, with plastic steering wheel and perspex windscreen, plastic flashing lamp mounted to hood, with battery located underneath dashboard. Solid steel painted balloon wheels with inset rubber tires. Complete with original period shipping carton. £200-£250; $300-$375.

Tri-ang Sharna Ware PLC Rolls-Royce Corniche convertible, produced from the early 1980s through the early 1990s. All-plastic monocoque with dummy drop-down top, molded windscreen with perspex windsheld. Spray chromed radiator grill with rubber "Spirit of Ecstasy" (flying lady) mascot on top. Working twin headlights and side with rear indicators and stop lights. Authentic dashboard with turn-key operation for lights and horn. Two versions were produced, one pedal powered and the other powered by a 12-volt electricity through registrations RRM 1 to RRM 6 in red, white, and blue. All were retailed with miniature road fund license (tax disc) windscreen sticker and owner's log book. Mainly retailed in top London department stores. The very last batch was purchased by C.A.R.S. of Brighton in 1991. Value for pedal version in good condition with no parts missing £350-£400 ($525-$600); the electric version £450-£500 ($675-$750).

Jaguar XK120, faithfully reproduced half-scale size in fiberglass, with working headlamps and red leather upholstery. Alloy bumpers and trim, scaled down wheels, proper pneumatic treaded tires, and alloy hubcaps. Seen at the Paris Retromobile classic car show and exhibition in 1995.

BMW quarter-scale version of the 1936 328 Sportswagen, with a Honda two-stroke petrol engine in the rear. Authentic dashboard with working dials and switches for horn, headlamps, and engine operation. Chrome grill and accessories, scale wheels with rubber treaded pneumatic tires, and twin hood straps. Features glass windshield mounted with spotlight.Upholstered in red leather. Value undeclared. *Courtesy of Cheyne BMW showroom, Parsons Green, Fulham, London, U.K.*

Tri-ang (Lines Bros. Ltd.) Supplement

The following pedal car illustrations are relevant reproduced pages taken from original sales catalogues and brochures promoting all pedal cars, bicycles, toys, prams, and dollhouses manufactured by The Lines Bros. group between the years 1955 through to 1980. After this period the majority of pedal car production was produced almost wholly in plastic with few metal parts, and the range was considerably reduced. The company was taken over by the French concern of Superjouet in 1990 and a few pedal cars are now produced with the Tri-ang trademark under the banner of Superjouet (U.K.) Ltd. The author is kindly indebted to Superjouet for permission to produce this information.

The monochrome studio photos were commissioned by the Tri-ang Co. in the 1960s for their in-house reference folder.

Thistle (Canada) No. 8030 Racing car.

Tri-ang TM0131 E-Type Jaguar with motor roar unit

Tri-ang TM0122 Junior Grand Prix Racer with motor roar unit

Tri-ang MOTORS — ENGINEERED FOR STRENGTH AND LIGHT WEIGHT

TRI-ANG " EIGHTY " B
(6001)

TRI-ANG " EIGHTY " B PEDAL MOTOR (6001)

This is the finest children's motor yet made. It is beautifully finished in super gloss enamel and brilliant chrome. The pressed steel body is a particularly fine job.

SIZE	48½ ins. (123 cms.) × 17½ ins. (45 cms.) wide. For ages 3 to 7 years.	HORN	Mechanical (or electric at extra cost).
BODY	Welded, heavy gauge pressed steel. No sharp edges. CHROMED decor.	WHEELS	10 ins. dia. balloon disc.—rubber tyred. Large CHROMED Caps.
BUMPERS	Pressed steel, mounted on rubber buffers, front and rear. CHROMED.	REAR AXLE	Fitted four ball-bearings.
		PEDALS	Adjustable for leg length.
RADIATOR GRILLE	Pressed steel. CHROMED.	FINISH	High gloss enamel in sparkling colours. BRIGHT PARTS CHROMED.
WINDSCREEN	Pressed steel. CHROMED.		
STEERING WHEEL	8½ ins. dia. with plated spring spokes.	PACKING	1 per carton for Export.

Page One

Tri-ang EXCLUSIVE DESIGNS OF MODERN MOTORS

TRI-ANG "SIXTY" FIRE CHIEF (6003)

TRI-ANG "EIGHTY" A (6002)

TRI-ANG "EIGHTY" A (6002)

SIZE	48½ ins. (123 cms.) × 17½ ins. (45 cms.) wide. For ages 3–7 years.
BODY	Welded heavy-gauge pressed steel.
BUMPERS	Front and Rear. Pressed steel—mounted on rubber buffers.
WHEELS	10 in. dia. balloon disc—rubber tyred.
BRAKE	Hand operated.
REAR AXLE	Fitted FOUR BALL-BEARINGS.
PEDALS	Adjustable for leg length—rubber foot pads.
FINISH	High gloss enamel with White contrast.
PACKING	1 per carton for Export.

TRI-ANG "SIXTY" FIRE CHIEF (6003)

SIZE	40½ ins. (102 cms.) × 17½ ins. (45 cms.) For ages 3–7 years.
BODY, RADIATOR GRILLE AND WINDSCREEN	in heavy-gauge pressed steel.
BUMPER	Front only. Pressed steel—mounted on rubber buffers.
WHEELS	8 in. dia. balloon disc—rubber tyred.
REAR AXLE	Fitted FOUR BALL-BEARINGS.
PEDALS	Adjustable for leg length—rubber foot pads.
FINISH	High gloss Red enamel with White and Yellow contrast.
PACKING	1 per carton for Export.

Page Two

Tri-ang CAREFULLY CONSTRUCTED ALL-STEEL MOTORS

TRI-ANG "THIRTY" A (6006)

TRI-ANG "THIRTY" B (6007)

TRI-ANG "FORTY" (6008)

TRI-ANG "THIRTY" A (6006)

SIZE	35 ins. (89 cms.) × 14 ins. (36 cms.). For ages 2–5½ years.
BODY	Welded, heavy-gauge pressed steel.
WINDSCREEN	Pressed steel.
WHEELS	6½ ins. dia. disc—with ¾ in. rubber tyres.
FINISH	High gloss enamel in sparkling colours with White contrast.
PACKING	1 per carton.

TRI-ANG "THIRTY" B (6007)

SIZE	35 ins. (89 cms.) × 14 ins. (36 cms.). For ages 2–5½ years.
BODY	Welded, heavy-gauge pressed steel.
WINDSCREEN	Pressed steel.
WHEELS	8 ins. dia. disc—with 1 in. rubber tyres.
FINISH	High gloss enamel in sparkling colours with White contrast. Chromed arrows on sides.
PACKING	1 per carton.

TRI-ANG "FORTY" (6008)

SIZE	40 ins. (102 cms.) × 17⅞ ins. (46 cms.). For ages 3–6 years.
BODY	Welded, heavy gauge pressed steel.
WINDSCREEN	Pressed steel.
WHEELS	8 ins. dia. balloon steel disc—rubber tyres.
PEDALS	Fitted rubber pads.
FINISH	High gloss enamel in sparkling colours with White contrast.
PACKING	1 per carton for Export.

Page Four

Tri-ang BALL-BEARING MOTORS FOR EASY PEDALLING

TRI-ANG "SIXTY" A (6004)

TRI-ANG "SIXTY" B (6005)

TRI-ANG "SIXTY" A (6004)

SIZE	40½ ins. (102 cms.) × 17½ ins. (45 cms.) For ages 3–7 years.
BODY	Welded, heavy-gauge pressed steel.
BUMPER	Front only. Pressed steel—mounted on rubber buffers.
WHEELS	8 in. dia. balloon disc—rubber tyred.
REAR AXLE	Fitted FOUR BALL-BEARINGS.
PEDALS	Adjustable for leg length—rubber foot pads.
FINISH	High gloss enamel in sparkling colours.
PACKING	1 per carton for Export.

TRI-ANG "SIXTY" B (6005)

SIZE	40½ ins. (102 cms.) × 17½ ins. (45 cms.) For ages 3–7 years.
BODY	Welded, heavy gauge pressed steel.
BUMPER	Front only. Pressed steel—mounted on rubber buffers. CHROMED.
WHEELS	8 in. dia. balloon disc—rubber tyred.
REAR AXLE	Fitted FOUR BALL-BEARINGS.
PEDALS	Adjustable for leg length—rubber foot pads.
FINISH	High gloss enamel in sparkling colours. CHROMED Radiator grille, lamps, bumper, windscreen and hub caps.
PACKING	1 per carton for Export.

Page Three

Tri-ang MOTORS—BY THE WORLD'S LARGEST TOY MANUFACTURERS

TRI-ANG RACER "SUPER-EIGHT" (PNEUMATIC) (6009)

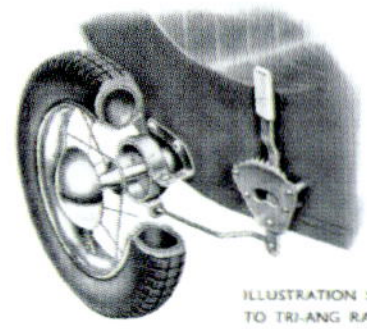

ILLUSTRATION SHOWS BRAKE UNIT FITTED TO TRI-ANG RACER "SUPER-EIGHT"

TRI-ANG RACER "B" (6010)

TRI-ANG RACER "SUPER-EIGHT" (PNEUMATIC) (6009)

SIZE	50 ins. (127 cms.) × 25 ins. (64 cms.) For ages 4–7 years.
BODY	Welded, heavy-gauge pressed steel.
WINDSCREEN	Moulded plastic.
WHEELS	12½ in. dia. tangent spoke.
TYRES	2¼ in. Dunlop pneumatic with Schraeder valves
BRAKE	External—hand operated. Ratchet controlled.
REAR AXLE	Fitted FOUR BALL-BEARINGS.
PEDALS	Adjustable for leg length—rubber foot pads.
FINISH	High gloss enamel. White, Red, Green or Blue, with BRIGHT PARTS CHROMED.
PACKING	1 per carton for Export.

TRI-ANG RACER "B" (6010)

SIZE	48½ ins. (123 cms.) × 20 ins. (51 cms.) For ages 4–7 years.
BODY	Welded, heavy-gauge pressed steel.
WINDSCREEN	Moulded plastic.
WHEELS	10 in. dia. resilient jointless rubber tyres on 6 in. dia. die-cast hubs.
BRAKE	External—hand operated.
REAR AXLE	Fitted FOUR BALL-BEARINGS.
PEDALS	Adjustable for leg length—rubber foot pads.
FINISH	High gloss enamel in sparkling colours. BRIGHT PARTS CHROMED.
PACKING	1 per carton for Export.

Page Five

Tri-ang BEAUTIFULLY FINISHED CARS FOR CHILDREN

TRI-ANG RACER "A" (6011)

PHOTOGRAPH OF BRAKE FITTED TO TRI-ANG RACER "B" AND RACER "A"

TRI-ANG RACER "X" (6012)

TRI-ANG RACER "A" (6011)

SIZE	48½ ins. (123 cms.) × 20 ins. (51 cms.). For ages 4–7 years.
BODY	Welded, heavy-gauge pressed steel.
WINDSCREEN	Moulded plastic.
WHEELS	10 in. dia. balloon disc—rubber tyred.
BRAKE	External—hand operated.
REAR AXLE	Fitted FOUR BALL-BEARINGS.
PEDALS	Adjustable for leg length—rubber foot pads.
FINISH	High gloss enamel in sparkling colours with White contrast.
PACKING	I per carton for Export.

TRI-ANG RACER "X" (6012)

SIZE	48½ ins. (123 cms.) × 20 ins. (51 cms.). For ages 4–7 years.
BODY	Welded, heavy-gauge pressed steel.
WHEELS	10 in. dia. balloon disc—rubber tyred.
REAR AXLE	Fitted FOUR BALL-BEARINGS.
PEDALS	Adjustable for leg length—rubber foot pads.
FINISH	High gloss enamel in sparkling colours with White contrast.
PACKING	I per carton for Export.

Tri-ang THE FINEST NAME IN CHILDREN'S MOTORS

TRI-ANG "SIXTY" STATION WAGON
(6013)

TRACTOR TRAILER
(6014/2)

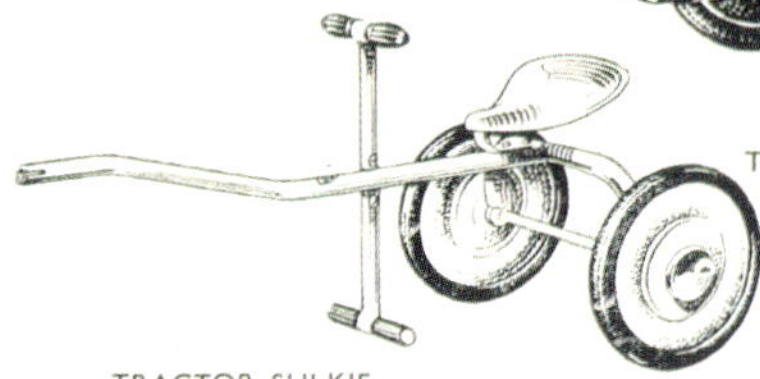

TRACTOR SULKIE
(6014/6)

TRACTOR
BULLDOZER SHIELD
(6014/5)

TRI-ANG "MAJOR" PEDAL TRACTOR
(6014)

TRI-ANG "SIXTY" STATION WAGON (6013)

SIZE	$40\frac{1}{4}$ ins. (102 cms.) × $17\frac{1}{2}$ ins. (45 cms.). For ages 3 to 7 years.
BODY	Welded, heavy gauge, pressed steel. Large rear goods compartment with opening tailboard.
RADIATOR GRILLE	CHROMED pressed steel.
WINDSCREEN	Pressed steel.
BUMPER	Front. Pressed steel—mounted on rubber buffers.
WHEELS	8 ins. dia. balloon disc—rubber tyred.
REAR AXLE	Fitted FOUR BALL-BEARINGS.
PEDALS	Adjustable for leg length—rubber foot pads.
FINISH	High gloss enamel in sparkling colours with rear section in Natural Wood colour. Bright parts CHROMED.
PACKING	I per carton for Export.

TRI-ANG "MAJOR" PEDAL TRACTOR (6014)

SIZE	35 ins. (89 cms.) long. For ages 3 to 7 years.
BODY	Pressed steel and welded tubular steel construction.
WHEELS	$10\frac{1}{2}$ ins. dia. front, 14 ins. dia. rear. Steel disc. Ball-bearing front. "Oversize" Balloon type rubber tyres.
REAR AXLE	CRANK DRIVE. Mounted on FOUR BALL-BEARINGS.
ACCESSORIES	(At extra cost.) Rake (6014/1) Trailer (6014/2), Roller (6014/3), Plough (6014/4), Bulldozer (6014/5), Sulkie (6014/6).
FINISH	High gloss enamel with contrast wheels and decor.
PACKING	I per carton for Export.

TRI-ANG "MINOR" PEDAL TRACTOR (6015)

Similar in design to the larger "MAJOR" model above but with lighter frame, 8 ins. (front) and 10 ins. (rear) balloon disc rubber tyred wheels and no mudguards.

Page Seven

Tri-ang PEDAL MOTORS LEAD THE WORLD IN QUALITY AND VALUE

TRI-ANG JEEP
(6016)

TRI-ANG FIRE ENGINE
(6017)

TRI-ANG JEEP (6016)

SIZE	40 ins. (102 cms.) × 17½ ins. (45 cms.) wide. For ages 4–7 years.
CHASSIS	Welded, tubular steel.
BODY	Pressed steel.
WINDSCREEN	Folding type.
WHEELS	10 in. dia. balloon steel disc—rubber tyred.
BUMPERS	Pressed steel.
REAR AXLE	Fitted FOUR BALL-BEARINGS.
PEDALS	Fitted rubber pads.
ACCESSORIES	Spare wheel and Jerry can.
FINISH	High gloss enamel—Military Khaki, with White insignia, numbers, etc. Alternatively in sparkling colours with lined contrast.
PACKING	I per carton for Export.

TRI-ANG FIRE ENGINE (6017)

SIZE	Body: 40 ins. (102 cms.) × 17½ ins. (45 cms.) wide. For ages 4–7 years.
CHASSIS	Welded, tubular steel.
BODY	Pressed steel.
WHEELS	10 in. dia. balloon steel disc—rubber tyred.
BUMPERS	Pressed steel.
REAR AXLE	Fitted FOUR BALL-BEARINGS.
PEDALS	Fitted rubber pads.
ACCESSORIES	Automatic fire bell; tubular steel extensible ladders; Jerry can and spare wheel.
FINISH	High gloss enamel. Red with Yellow contrast.
PACKING	I per carton for Export.

Page Eight

Tri-ang

ATTRACTIVELY PRICED MOTORS FOR QUICK SALES

TRI-ANG JEEP STATION WAGON (6019)

TRI-ANG DAIMLER SPORTS "8" (PNEUMATIC) (6020)

TRI-ANG JEEP STATION WAGON (6019)

SIZE	46 ins. (117 cms.) × 17½ ins. (45 cms). wide. For ages 4-7 years.
CHASSIS	Welded, tubular steel.
BODY	Pressed steel. Rear goods compartment, 14 ins.×9 ins.×7 ins.
WINDSCREEN	Folding type.
WHEELS	10 in. dia. balloon steel disc—rubber tyred.
BUMPER	Front only—Pressed steel.
REAR AXLE	Fitted FOUR BALL-BEARINGS.
PEDALS	Fitted rubber pads.
ACCESSORIES	Spare wheel. Jerry can.
FINISH	High gloss enamel in sparkling colours. Bright parts CHROMED.
PACKING	1 per carton.

TRI-ANG DAIMLER SPORTS "8" (6020)

SIZE	54 ins. (137 cms.) × 24½ ins. (62 cms.) wide. For ages 5-8 years.
BODY	Light weight, coach built. Side door.
SEAT	Upholstered leather cloth.
MUDGUARDS	Pressed steel, wide-section cycle type.
WHEELS	12½ in. dia. tangent spokes.
TYRES	2½ in. Dunlop pneumatic. Motor type. With Schraeder valves.
REAR AXLE	Fitted FOUR BALL-BEARINGS. Independent springing.
PEDALS	Adjustable for leg length—rubber foot pads.
FINISH	High gloss enamel in sparkling colours. Bright parts CHROMED.

Page Nine

Tri-ang

MOTORS—THERE IS NO BETTER VALUE OBTAINABLE

TRI-ANG DUKE No. 1 (6021)

TRI-ANG DUKE No. 2 (6022)

TRI-ANG DUKE FIRE ENGINE (6023)

TRI-ANG DUKE No. 1 (6021)

SIZE	30 ins. (76 cms.) × 15 ins. (38 cms.) wide. For ages 2–4 years.
BODY	Pressed steel.
WHEELS	6½ in. dia. steel disc—rubber tyred.
REAR AXLE	Plain bearing.
FINISH	High gloss enamel with contrast lining.
PACKING	(Home) 1 per carton. (Export) 2 per carton.

TRI-ANG DUKE FIRE ENGINE (6023)

SIZE	31 ins. (79 cms.) × 15 ins. (38 cms.) wide. For ages 2–4 years.
BODY	Pressed steel.
WHEELS	8 in. dia. steel disc—1 in. rubber tyres.
REAR AXLE	Plain bearing.
ACCESSORIES	Twin ladders. Jerry can. Fire bell.
FINISH	High gloss enamel. Red with Yellow contrast.
PACKING	(Home) 1 per carton. (Export) 2 per carton.

TRI-ANG DUKE No. 2 (6022)

SIZE	32 ins. (81 cms.) × 15 ins. (38 cms.) wide. For ages 2–4 years.
BODY	Pressed steel.
WHEELS	8 in. dia. steel disc—1 in. rubber tyres.
REAR AXLE	Plain bearing.
ACCESSORY	Jerry can.
FINISH	High gloss enamel with contrast lining.
PACKING	(Home) 1 per carton. (Export) 2 per carton.

Page Ten

Tri-ang TM0210 Countryman

Tri-ang (ref. unknown) Weekender

Tri-ang
Regd. Trade Mark
Toys
THE LARGEST TOY MANUFACTURING ORGANISATION IN THE WORLD
LINES BROS GROUP
CATALOGUE NO. 72 1955/6

Tri-ang

Regd Trade Mark

TRI-ANG EIGHTY (6002)

Length 48½″ (123 cms.). For ages 3 to 7 years. Improved for 1955, this popular, high-class model features welded, heavy-gauge, pressed-steel body with opening rear boot ; sprung, 3-spoke steering wheel ; front and rear chromed bumpers ; 10″ dia. balloon disc wheels with rubber tyres ; hand-operated brake ; crank drive with ball-bearing rear axle ; and adjustable pedals for leg length. Accessories : Dummy gear shift lever ; mechanical windscreen wiper ; trafficators ; rubber aeroplane mascot and A.A. Badge. High gloss enamel and chrome finish. Packed 1 per carton.

TRI-ANG T 60C TIPPER (8003)

TRI-ANG T/60 CRASH WAGON (7003)

Length 51″ (130 cms.). For ages 3 to 7 years. Really lifts and tows another pedal car. Specification : Aero-styled pressed-steel body with welded extension ; strong, pressed-steel crane with ratchet controlled hoisting gear and screw-jack operated jib ; sprung, 3-spoke steering wheel ; 8″ dia balloon disc wheels with rubber tyres ; fully ball-bearing axle ; adjustable pedals for leg length ; and mechanical windscreen wiper. High gloss enamels with bright parts chromed. Packed 1 per carton.

TRI-ANG T/60C TIPPER (8003)

Length 49″ (125 cms.). For ages 3 to 7 years. Specification : Generally similar to T/60 Crash Wagon above but fitted with lever-operated hopper which can carry a big load. Beautifully finished in chrome and enamel. Packed 1 per carton.

Tri-ang

Regd Trade Mark

TRI-ANG T 60 FIRE CHIEF (6003)

Length 41½″ (105 cms.). For ages 3 to 7 years. Basically similar to the T/60 Police Car, this excellent model comes in brilliant Red enamel with yellow markings, and fire bell. Packed 1 per carton.

TRI-ANG T 60 POLICE CAR (8004)

Length 41½″ (105 cms.). For ages 3 to 7 years. Fully equipped for the young " policeman " this pedal motor is finished in Police Blue enamel with chrome fittings. It features an aero-styled pressed-steel body with ball-bearing crank drive ; 8″ dia. balloon wheels with rubber tyres ; and has adjustable pedals for leg length. Accessories : Twin driving mirrors ; dummy twin loudspeakers ; two-tone horn ; radio mast ; a loud clanging bell ; dummy gear shift lever and mechanical windscreen wiper. Packed 1 per carton.

TRI-ANG T 60 STATION WAGON (6013)

Length 51″ (130 cms.). For ages 3 to 7 years. Generally similar in basic design to the T/60 Police Car above, this model features a large capacity rear compartment with opening tailboard. Accessories : Mechanical windscreen wiper. Packed 1 per carton.

Tri-ang
Regd. Trade Mark

TRI-ANG T 30B LIGHTNING (6007)

Length 35" (89 cms.). For ages 2 to 5½ years. Although comparatively inexpensive, this model is exceptionally strong and will give first-class service. The Lightning features a welded, pressed-steel body with chromed radiator grille and decor; 8" dia. disc wheels with 1" rubber tyres; and pedal operated crank drive. Accessory: Two-tone horn. Brilliantly finished in sparkling enamel with bright parts chromed. Packed 1 per carton.

TRI-ANG T 30A THUNDERBOLT (6006)

Specification as the T/30B Lightning, but fitted with 6½" dia. wheels with ½" tyres and not fitted with two-tone horn. High gloss enamel finish with bright parts chromed. Packed 1 per carton.

TRI-ANG T 30C TIPPER (8005)

Length 43½" (110 cms.). For ages 3 to 5 years. A new model with universal appeal. Basically the same specification as the T/30B Lightning, the model has a large capacity hopper operated by lever from driver's seat. Beautifully finished with hopper in contrast colour and bright parts chromed. Packed 1 per carton.

Page Seven

Tri-ang
Regd. Trade Mark

TRI-ANG "MIDGET" TRACTOR (8008)

Length 30" (76 cms.). For ages 2 to 5 years. A splendid model beautifully finished with realistic "engine." Strongly made with 1" tubular steel chassis surmounted with pressed-steel boiler with chromed end plates and rubber exhaust pipe. Fool-proof link action steering for turning in small radius. 8" dia. front and 10" dia. rear artillery type wheels with NYLOMATIC bearings, 1" jointless tyres and chromed caps on rear wheels. Non-slip rubber pedals. Brilliantly finished in Red with Yellow wheels and bright parts chromed. Packed 1 per carton.

TRI-ANG "MAJOR" PEDAL TRACTOR (6014)

Length 35" (89 cms.). For ages 3 to 7 years. An established favourite, this extra strong model has a first-class specification. The chassis and body are of pressed-steel and tubular steel construction and the front forks have a ball-bearing fitment. It has 10" dia. front and 11½" dia. rear balloon tyres and a ball-bearing crank drive. High gloss enamel finish with bright parts chromed. ACCESSORIES as illustrated below at extra cost. Packed 1 per carton for export.

TRACTOR TRAILER (6014.2)
TRACTOR RAKE (6014.1)
TRACTOR SULKIE (6014.6)
TRACTOR BULLDOZER SHIELD (6014.5)

TRI-ANG "SUPER" TRACTOR (7611)

Similar in size and construction to the "Major" Tractor, this model incorporates the following features: (1) Chain drive, ball-bearing transmission. (2) 9" dia. front disc wheel with 1½" super cushion tyre. (3) 10½" dia. rear disc wheels with oversize treaded super-cushion tyres. (4) Pressed-steel crane with elevation control for jib and ratchet controlled hoist.

Page Eleven

Tri-ang
Regd. Trade Mark

TRI-ANG RACER "SUPER EIGHT" (PNEUMATIC) (6009)

Length 50" (127 cms.). For ages 4 to 7 years. Features of this distinctive model are the streamlined auto-steel body with transparent plastic racing shield; fully ball-bearing crank drive; adjustable pedals; 12½" dia. tangent spoke wheels; Dunlop pneumatic tyres with Schraeder valves; hand-operated, ratchet controlled band brake (see inset) and the brilliant chromed and enamel finish. Packed 1 per carton.

TRI-ANG SILVERSTONE RACER (8007)

Length 34" (86 cms.). For ages 3 to 5 years. New for 1955 this superb model incorporates an enclosed chain drive and ball-bearing rear axle in its magnificent specification. It has a pressed-steel body with an integral exhaust manifold and pipe; clear plastic racing shield; chromed grille; 8" dia. artillery type wheels with 1" jointless tyres; and it is beautifully finished in high gloss enamel and chrome. Packed 1 per carton.

TRI-ANG RACER "A" (6011)

Length 48½" (123 cms.). For ages 4 to 7 years. Generally similar to the "Racing Eight", this model is now fitted with 10" dia. white-walled balloon wheels with rubber tyres; 3-spoke sprung steering wheel; hand brake operating on rear wheel and chromed radiator grille, exhaust pipe and hub caps. Packed 1 per carton.

Page Ten

Tri-ang
Regd. Trade Mark

TRI-ANG JEEP (6016)

Length 40" (102 cms.). For ages 4 to 7 years. An improved version of our world-famous model. It incorporates a welded, pressed-steel body with bumpers; sprung, 3-spoke steering wheel; ball-bearing crank drive; rubber pad pedals adjustable for leg length; and 8" artillery type wheels with NYLOMATIC bearings and 1" jointless tyres. Accessories: Spare wheel and Jerrycan. Finished in a choice of two colours—Khaki with Military Police and Army markings or Air Force Blue with R.A.F. markings. Bright parts chromed. Packed 1 per carton.

TRI-ANG DUKE FIRE ENGINE (6023)

Length 31" (79 cms.). For ages 2 to 4 years. Designed particularly for the very young "fireman" this steel-built model has an extra strong welded body; pedal driven crank drive; 8" dia. steel disc wheels with 1" jointless tyres, twin hardwood ladders and a firebell. Beautifully finished in Red with contrast Yellow markings, etc. Packed 1 per carton (Home), 2 per carton (Export).

TRI-ANG JEEP FIRE ENGINE (6017)

Generally similar in design and specification to the Jeep (6016) this super model is fitted with chromed firebell which automatically rings when the motor is pedalled, and 2 tubular steel, extensible ladders. Brilliant Red finish with Yellow ladders, wheels, markings, etc. Bright parts chromed. Packed 1 per carton.

Page Eight

Tri-ang
Regd. Trade Mark

TRI-ANG PRINCE (7020)

Identical in design and specification to the Royal Prince on right, except that it is fitted with an enamelled radiator front. Packed 1 per carton (Home), 2 per carton (Export).

TRI-ANG ROYAL PRINCE (8006)

Length 31" (79 cms.). For ages 2 to 4 years. Latest addition in the inexpensive TRI-ANG range of pedal motor cars, this model is remarkable for its superb chrome and enamel finish. It has a welded, pressed-steel body; pedal driven crank drive; 8" dia. steel disc wheels with 1" jointless tyres and a two-tone horn. Packed 1 per carton (Home), 2 per carton (Export).

TRI-ANG DUKE No. 1 (6021)

Length 30½" (78 cms.). For ages 2 to 4 years. Although the lowest priced model in the TRI-ANG range, the "DUKE" No 1 features a strong, pressed-steel body; pedal driven crank drive; 7" dia. steel disc wheels with rubber tyres. Beautifully finished in high gloss enamel. Packed 1 per carton (Home), 2 per carton (Export).

Tri-ang
Regd. Trade Mark

TRI-ANG NEPTUNE (8002)

Length 40" (102 cms.). For ages 4 to 7 years. A magnificent new model featuring enclosed chain drive transmission. It has an aero-styled, pressed-steel body with chromed bumpers, radiator grille, etc. 8" dia. artillery type wheels with NYLOMATIC bearings and 1" jointless tyres; and fully ball-bearing rear axle. Accessories: Rubber aeroplane mascot, mechanically operated windscreen wiper, dummy gear shift lever and two-tone horn. Gleaming chrome and enamel finish. Packed 1 per carton.

TRI-ANG NORTH STAR (7000)

The only Pedal Car with SHAFT DRIVE and 2-GEAR TRANSMISSION.

Length 40" (102 cms.). Suitable for ages 4 to 7 years. This superb model has a special aero-styled, pressed-steel body with chromed radiator grille, head lamps, windscreen and bumpers. It is propelled by cycle type pedals via bevel gears and shaft drive to the gear box, which provides two forward and reverse speeds and which is operated by chromed gear lever on shaft casing. Fully ball-bearing. 8" dia. artillery type wheels with NYLOMATIC bearings and 1" jointless tyres. Accessories include brake, 2-tone horn, mechanically operated windscreen wiper, manually operated trafficators, rubber aeroplane mascot and A.A. Badge, super gloss enamel and chrome finish. Packed 1 per carton.

TRI-ANG JUPITER (7601)

Length 40" (102 cms.). For ages 4 to 7 years. Generally similar to the Neptune, the Jupiter is fitted with pedal-crank action drive, ball-bearing rear axle and 8" dia. artillery type wheels with NYLOMATIC bearings and 1" jointless tyres. Accessories include two-tone horn, "dummy" gear shift lever, rubber mascot, A.A. Badge and mechanical windscreen wiper. Packed 1 per carton.

Page Three

Tri-ang

Regd. Trade Mark

TRI-ANG CENTURION (7001)

The finest equipped pedal motor in the world. Specification : Length 48½" (123 cms.). For ages 4 to 7 years. Welded, heavy-gauge, pressed steel body with chromed radiator grille and bumpers. Ball-bearing crank drive. Adjustable pedals. 9½" dia. large section cushion tyred wheels with chromed caps. Accessories : dashboard mounted "Radio" which plays a popular tune ; electric horn operated from button on sprung steering wheel ; electric headlamps and red tail light with on/off switch ; opening boot which houses batteries, etc. ; safety hand brake ; mechanically operated windshield wiper, trafficators ; dummy gear shift lever, A.A. badge and rubber aeroplane mascot. It is a beautifully designed car with a brilliant chrome and enamel finish. Packed 1 per carton.

TRI-ANG CENTURION (7001)

TRI-ANG GRAND PRIX ELECTRIC RACER (8001)

TRI-ANG GRAND PRIX ELECTRIC RACER (8001)

Length 53¾" (137 cms.). For ages 4 to 7 years. A truly magnificent children's car, powered by a precision made electric motor operated by a chromed lever at side. Forward, neutral and reverse gears. The streamlined body is constructed from welded heavy-gauge pressed steel and has a hinged plate at rear for access to the 12 volt accumulator (not supplied). The tangent spoke wheels are 12½" dia. overall and are fitted with Dunlop 2¼" pneumatic tyres with Schraeder valves. Fully ball-bearing. Hand operated, ratchet controlled, hand brake. High gloss enamel and scintillating chrome finish. Packed 1 per export carton.

Page Two

Tri-ang

Regd. Trade Mark

TRI-ANG T 60 ARIEL (7007)

Length 41¼" (105 cms.). For ages 3 to 7 years. A well-equipped, moderately priced, pedal motor incorporating an aero-styled auto-steel body with chromed radiator grille and bumper ; crank drive with ball-bearing rear axle ; rubber pedals adjustable for leg length ; 8" dia. balloon disc wheels with rubber tyres. Accessories : Rubber aeroplane mascot ; two-tone horn ; mechanical windscreen wiper and dummy gear shift lever. High gloss enamel finish with bright parts chromed. Packed 1 per carton.

TRI-ANG T 70 CRUSADER (7606)

Length 41¼" (105 cms.). For ages 3 to 7 years. Generally similar to the T/60 Ariel but fitted with 8" dia. artillery type wheels with NYLOMATIC bearings and 1" tyres and plated two-tone horn. Packed 1 per carton.

TRI-ANG T 50 METEOR (7009)

Length 41" (104 cms.). For ages 3 to 7 years. Generally similar to the "T 60" Ariel but not fitted with windscreen wiper, two-tone horn, dummy gear shift lever, mascot or front bumper. Packed 1 per carton.

Page Six

Tri-ang Go-kart (ref. no. unknown)

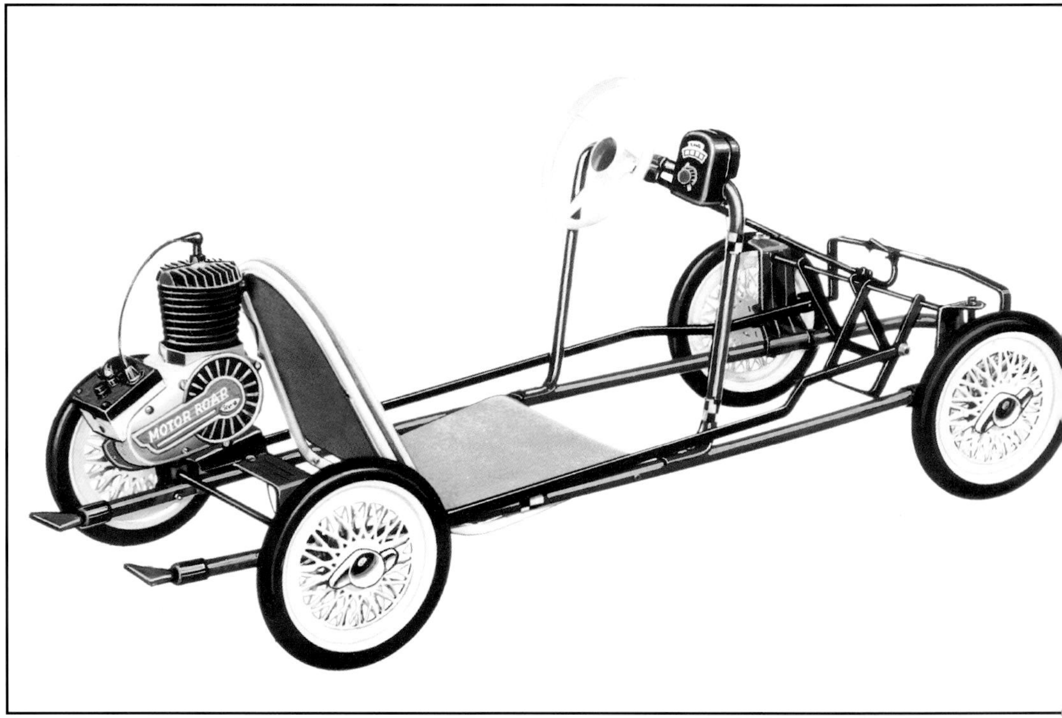

Tri-ang TW0911 Go-kart 'A' with motor roar unit

Tri-ang
Regd. Trade Mark
Toys
1961

TRI-ANG
FORMULA 1
RACING CAR

TRI-ANG
GRAND PRIX
RACING CAR

Tri-ang Pedal Motors

Regd Trade Mark

TRI-ANG FORMULA I RACING CAR

Glass-fibre body, 56" long on tubular chassis, with an adjustable leg measurement of between 16"-19". 12" dia. laced wheels in brilliant chrome with imitation "knock-off" caps; fitted with chromed exhaust pipe, and handbrake.

TRI-ANG GRAND PRIX RACING CAR

47" pressed-steel body on tubular chassis, with chromed exhaust tube and new type steering wheel, transparent windscreen and 9" balloon wheels with imitation "knock-off" hub caps.

TRI-ANG M.G.S. (SUPER-CHARGED)

45" glass-fibre body moulded in one piece, on all-metal chassis, with chromed bumpers and grille. Electric head-lights and horn operated from dashboard. Toughened plastic wheels with chromed imitation "knock-off" hub caps. Transparent windscreen. Fitted 2 : 1 ratio drive giving faster speed with normal pedalling action.

TRI-ANG M.G.A. (STANDARD)

As M.G.S. Super-Charged, but with normal crank action.

TRI-ANG M.G.S.
(SUPER-CHARGED)

TRI-ANG M.G.A.
(STANDARD)

Page Eight

TRI-ANG GENERAL JEEP

40 pressed-steel body with bumpers. 8" balloon disc wheels with rubber tyres. Fitted with dummy telephone receiver, Jerrycan and spare wheel. Adjustable for 14"-17" leg measurement. Military markings and regimental flashes and "4 star" general flag.

TRI-ANG A.A. SCOUT CAR

Length 37". A welded pressed-steel body with ball-bearing drive, pedals adjustable for 14"-17" leg measurement. 8" dia. balloon disc wheels with rubber tyres and chromed caps. With electric lights.

Tri-ang

Regd. Trade Mark

Pedal Motors

TRI-ANG LAND ROVER

Welded pressed-steel body, pedal crank drive, adjustable for 14"-17" leg length. 8" dia. balloon disc wheels. Enamel finish. Length 36".

TRI-ANG JEEP

Length 40". Body of welded pressed steel with bumpers. 8" dia. balloon disc wheels with rubber tyres. Fitted with ball-bearing drive with pedals adjustable for 14"-17" leg measurement, spare wheel and Jerrycan. Military Police and Army markings.

Page Six

Tri-ang
Regd. Trade Mark

Pedal Motors

The 1961 range of Pedal Motor Cars is outstanding for design, authenticity and overall finish. This year's models include such features as streamlined bodies with opening bonnet and boot, dummy engines with "oil" and "water" tanks, and winking lights.

TRI-ANG MORRIS OXFORD

TRI-ANG WOLSELEY

TRI-ANG WOLSELEY

43" pressed steel body with 7" pressed-steel wheels with black tyres. Chromed radiator grille with Wolseley motif, white plastic bumpers and headlight shield. Combined pedal and seat back which is adjustable giving leg length of between 20"-25". Fitted with opening bonnet and boot. Electrically operated winking lights, radio, horn and dummy engine with "petrol", "oil" and water filling tank. Large capacity boot and softly padded plastic seat.

TRI-ANG M.G. MAGNETTE *(Not illustrated)*

Basic body design as for Wolseley, with chromed radiator white plastic bumpers and headlight shields. Opening bonnet and boot. Fitted with dummy engine only; electrically operated lights, radio and horn with combined adjustable pedal and seat back to give maximum leg length of 25".

TRI-ANG MORRIS OXFORD

43" pressed steel body with 7" pressed-steel wheels with black tyres. Chromed radiator grille of Morris Oxford design, white plastic bumpers and headlight shield. Combined pedal and seat back which is adjustable to leg length of 20"-25". Fitted with opening bonnet and boot, electrically operated lights and horn.

TRI-ANG AUSTIN A60 *(Not illustrated)*

Basic body design as for Morris Oxford, but fitted with enamelled wheels. Electrically operated lights, fitted with opening bonnet and boot. Adjustable pedal and seat back gives maximum leg length of 25".

Tri-ang
Regd. Trade Mark

TRI-ANG T.40B MONTE CARLO

TRI-ANG T.30C TIPPER

TRI-ANG T.30 AMBULANCE

TRI-ANG T.40B MONTE CARLO

Pressed-steel body, crank drive, suitable for 15" leg measurement. 7" dia. disc wheels, electric lights and streamlined horn. Finished in two-colour enamel and chrome. Length 36".

TRI-ANG T.30C TIPPER

New type pressed-steel body, 7" dia. wheels with chromed hub caps. Tipper operated by lever from driving seat. Crank drive, suitable for 15" leg measurement. High-gloss enamel finish. Length 43".

TRI-ANG T.30 AMBULANCE

New type pressed-steel body, 7" dia. wheels with chromed hub caps. Crank drive, suitable for 15" leg measurement. Detachable stretcher. Length 35".

TRI-ANG NODDY CAR

TRI-ANG ROYAL PRINCE

TRI-ANG COMET "B"

TRI-ANG DUKE No. 1

TRI-ANG COMET "A"

Tri-ang
Regd. Trade Mark

Pedal Motors

TRI-ANG NODDY CAR

Authentic model in pressed steel with balloon tyres, rubber mudguards and chromed radiator with starting handle, dummy headlights and two-tone horn. Adjustable for 14"-17" leg measurement. Body length 36".

TRI-ANG ROYAL PRINCE

Pressed-steel body, crank drive, 8" dia. disc wheels, with chromed hubs. Chrome and enamel finish. Length 30".

TRI-ANG DUKE No. 1

Strong pressed-steel body, pedal crank drive, 7" dia. disc wheels with 1" cushion tyres. Suitable for child with 16" leg measurement. High-gloss enamel finish. Length 30".

TRI-ANG COMET "B"

Pressed-steel construction with distinctive body markings. Plated radiator grille and electric headlights operating from dashboard. Pedal crank drive suitable for 15" leg length. Disc wheels with cushion rubber tyres. Length 35".

TRI-ANG COMET "A"

Heavy duty body with distinctive body markings. Adjustable pedal crank drive, suitable for 14"-15" leg measurement. Disc wheels with rubber tyres. Length 25".

Pedal Motors

TRI-ANG T.30 POLICE CAR

TRI-ANG T.30 FIRE CHIEF

TRI-ANG T.30A THUNDERBOLT

TRI-ANG T.30B LIGHTNING

TRI-ANG T.30 FIRE CHIEF

Pressed-steel construction. Length 35". Crank drive, suitable for 15" leg measurement. 6½" dia. disc wheels with chromed hub caps. Detachable ladders mounted on sides. Chromed bell on bonnet.

TRI-ANG T.30 POLICE CAR

New type steel body, pedal crank drive, suitable for 15" leg measurement. 7" dia. disc wheels, with chromed hub caps. Streamlined horn. Sparkling enamel and chrome finish. Length 35".

TRI-ANG T.30A THUNDERBOLT

Steel body, pedal crank drive, suitable for 15" leg measurement. 6" dia. disc wheels with chromed hub caps. High gloss enamel and chrome finish. Length 35".

TRI-ANG T.30B LIGHTNING

New type pressed-steel body, crank drive, suitable for 15" leg length. 7" dia. disc wheels with chromed hub caps. Streamlined horn. Sparkling enamel and chrome finish. Length 35".

TRI-ANG TRIUMPH HERALD "B"

TRI-ANG T.45C MIAMI

TRI-ANG T.45A

TRI-ANG T.45B

Tri-ang
Regd. Trade Mark

TRI-ANG T.45C MIAMI

[illegible]

TRI-ANG T.45B

[illegible]

TRI-ANG T.45A

[illegible]

TRI-ANG TRIUMPH HERALD "B"

[illegible]

TRI-ANG TRIUMPH HERALD "A"

[illegible]

Tri-ang
Regd. Trade Mark

Pedal Motors

TRI-ANG T.60 POLICE CAR

Length 40". Pressed-steel body with ball-bearing crank drive, adjustable for 13"-16" leg length. 8" dia. balloon wheels with rubber tyres. Fittings include electric headlights operating from dashboard, twin driving mirrors. Dummy loudspeakers, two-tone horn, radio mast, loud alarm bell, dummy gear lever and plastic-covered seat.

TRI-ANG T.65

Length 40". Steel body with chromed hubs and bumpers; crank drive, adjustable for 13"-16" leg length; and ball-bearing rear axle; 8" dia. balloon disc wheels with rubber tyres. Fittings include plastic covered seat, electric lights and new type electric horn. High-gloss duotone finish.

TRI-ANG T.50

Length 40". An attractive motor incorporating an all-steel body, with ball-bearing crank drive, adjustable pedals for 13"-16" leg length. 8" balloon disc wheels with rubber tyres. Glossy enamel finish.

TRI-ANG T.60 POLICE CAR

TRI-ANG T.65

TRI-ANG T.50

Tri-ang Pedal Motors

TRI-ANG SUPER GO-KART

48" body of strong tubular construction, suitable for 20" leg measurement. 12½" chromed laced wheels with solid balloon-type tyres. Steering operates on front wheels.

TRI-ANG JUNIOR GO-KART "B"

(Not illustrated)

Similar to Senior model above, but being 36" long overall. With high back seat, rubber fins to exhaust and rubber dummy engine. Pressed-steel wheels.

TRI-ANG SUPER TRACTOR

Chain drive, ball-bearing transmission. 9" dia. front disc wheel with 1¼" super-cushion tyre. 10¼" dia. rear disc wheels with oversize treaded solid cushion tyres. Pressed-steel crane with elevation control for jib and ratchet controlled hoist. Finished in bright contrasting colours. Length 38". Suitable for 18" inside leg measurement.

TRI-ANG MIDGET TRACTOR

This popular model beautifully finished. Constructed with a tubular and pressed-steel chassis and metal fuel tank fitted with chrome end plates and rubber exhaust pipe. Foolproof link action steering for turning in small ratios. Solid rubber pedals. 9" front and 10" rear disc wheels with nylomatic bearings and 1" cushion tyres. Overall length 30". Inside leg measurement 15".

TRI-ANG SUPERSONIC PLANE

Streamlined jet-type aeroplane in pressed steel and finished in flambuoyant colour. The swept-back delta type "Wings" fitted with red and green electric navigation lights together with red rear lights. Three-wheel "undercarriage" with the new type laced wheels and white cushion tyres.

Page Nine

Tri-ang TM0203 Police radio car

Tri-ang TM0070 Fire service truck

Tri-ang
Regd. Trade Mark
Toys
LBL 4242
1962

TRI-ANG FORMULA I RACING CAR
TRI-ANG FORMULA 1 RACING CAR
TRI-ANG GRAND PRIX RACING CAR
TRI-ANG GRAND PRIX RACING CAR
TRI-ANG M.G. (SUPER-CHARGED)
TRI-ANG SUPERSONIC PLANE
Brands Hatch
TRI-ANG M.G. (SUPER-CHARGED)
TRI-ANG SUPERSONIC PLANE

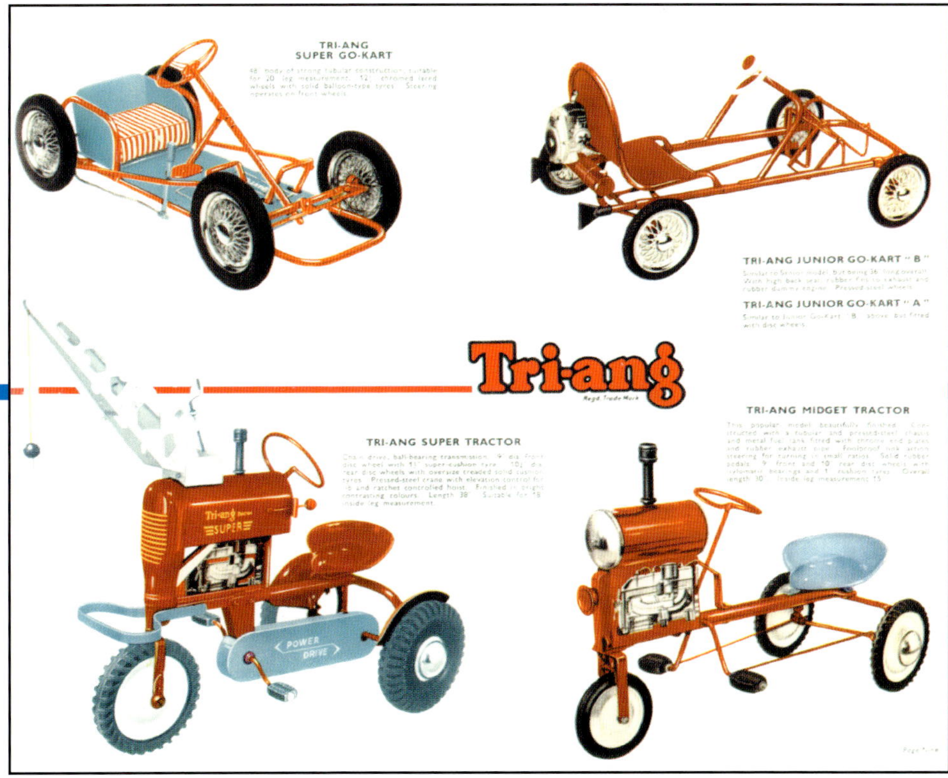
TRI-ANG SUPER GO-KART
TRI-ANG JUNIOR GO-KART "B"
TRI-ANG JUNIOR GO-KART "A"
Tri-ang
TRI-ANG SUPER TRACTOR
TRI-ANG MIDGET TRACTOR
POWER DRIVE

Tri-ang

Regd. Trade Mark

Pedal Motors

TRI-ANG GENERAL JEEP

40" pressed-steel body with bumpers. 8" red balloon disc wheels with rubber tyres. Fitted with dummy telephone receiver, Jerrycan and spare wheel. Adjustable for 14"-17" leg measurement. Military markings and regimental flashes. "4 star" General flag.

TRI-ANG A.A. SCOUT CAR

Length 37". A welded pressed-steel body with ball-bearing drive, pedals adjustable for 14"-17" leg measurement. 8" dia. balloon disc wheels with rubber tyres and chromed caps. With electric lights.

TRI-ANG LAND ROVER

Welded pressed-steel body, pedal crank drive, adjustable for 14"-17" leg length. 8" dia. balloon disc wheels. Enamel finish. Length 36".

TRI-ANG JEEP

Length 40". Body of welded pressed steel with bumpers. 8" dia. balloon disc wheels with rubber tyres. Fitted with ball-bearing drive with pedals adjustable for 14"-17" leg measurement, spare wheel and Jerrycan. Military Police and Army markings.

TRI-ANG MORRIS OXFORD

TRI-ANG WOLSELEY

Wooden Pedal Engine

TRI-ANG WOLSELEY

43" pressed-steel duotone body with 7" pressed-steel wheels with black tyres. Chromed radiator grille with Wolseley motif; white plastic over-riders and headlight shields and chromed bumpers. Combined pedal and seat back is adjustable giving leg length of between 20"-25". Fitted with opening bonnet and boot. Electrically operated lights, radio, horn and dummy engine with "petrol", "oil" and water filling tank. Large capacity boot and softly padded plastic seat.

TRI-ANG M.G. MAGNETTE *(Not illustrated)*

Basic body design as for Wolseley, with chromed radiator, white plastic over-riders and headlight shields with chromed bumpers. Opening bonnet and boot. Fitted with dummy engine only. Electrically operated lights, radio and horn with combined adjustable pedal and seat back to give maximum leg length of 25".

TRI-ANG MORRIS OXFORD

43" pressed-steel duotone body with 7" pressed-steel wheels with black tyres. Chromed radiator grille of Morris Oxford design, white plastic over-riders and headlight shields with chromed bumpers. Combined pedal and seat back which is adjustable to leg length of 20"-25". Fitted with opening bonnet and boot, electrically operated lights and horn.

TRI-ANG AUSTIN A60 *(Not illustrated)*

Basic body design as for Morris Oxford, but fitted with enamelled wheels, radiator and bumpers. Electrically operated lights. Fitted with opening bonnet and boot. Adjustable pedal and back gives maximum leg length of 25".

WOODEN PEDAL ENGINE "PRAIRIE CHIEF"

43" long overall, wooden construction throughout with crank drive. 4½" die cast front wheels and 6" dia. die cast rear wheels with white cushion tyres. Steering operated from lever inside cabin. As the engine is pedalled along, the imitation smoke chuffs up and down in the chimney. Fitted with toughened rubber "cow-catcher" and lamp. Chromed bell, fitted with pulling string. 14" leg length. Brilliantly finished in bright enamels.

Page Two

TRI-ANG NODDY CAR

TRI-ANG ROYAL PRINCE

TRI-ANG DUKE No. 1

TRI-ANG COMET "B"

Tri-ang

Regd. Trade Mark

Pedal Motors

TRI-ANG COMET "A"

TRI-ANG NODDY CAR

Authentic model in pressed steel with black body markings. Fitted with balloon disc wheels, rubber mudguards and chromed radiator with starting handle, dummy headlights and two-tone horn. Adjustable for 14"-17" leg measurement. Body length 36".

TRI-ANG ROYAL PRINCE

Pressed-steel body, crank drive. 8" dia. disc wheels, with chromed hubs. Chrome and enamel finish. Length 30".

TRI-ANG DUKE No. 1

Strong pressed-steel body, pedal crank drive, 7" dia. disc wheels with 1" cushion tyres. Suitable for child with 16" leg measurement. High-gloss enamel finish. Length 30".

TRI-ANG COMET "B"

Pressed-steel construction with distinctive body markings. Plated radiator grille and electric headlights operating from dashboard. Pedal crank drive suitable for 15" leg length. Disc wheels with cushion rubber tyres. Length 26".

TRI-ANG COMET "A"

Heavy duty body with distinctive body markings. Adjustable pedal crank drive, suitable for 14"-15" leg measurement. Disc wheels with rubber tyres. Length 26".

Page Seven

TRI-ANG T.40B MONTE CARLO

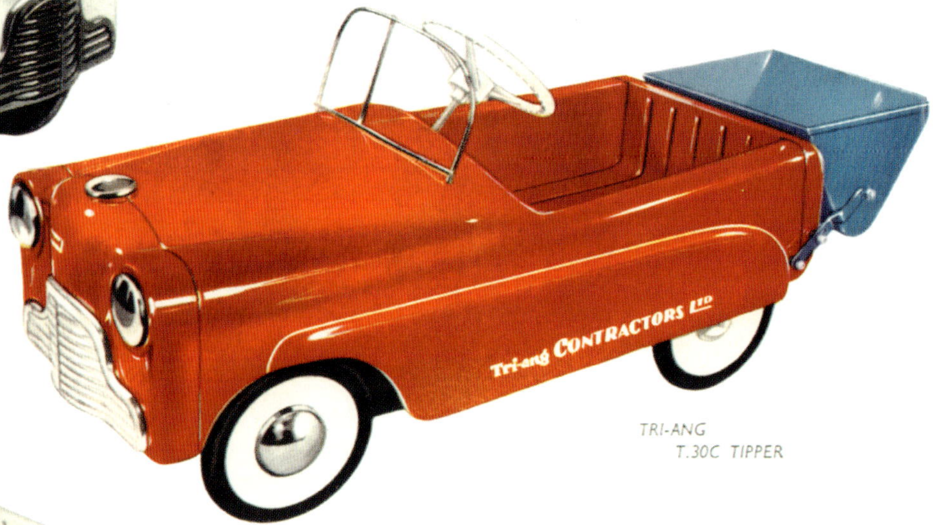

TRI-ANG T.30C TIPPER

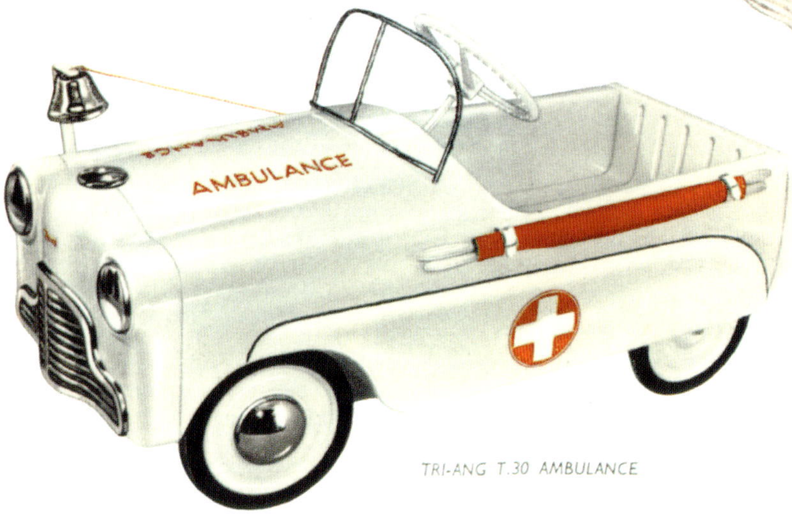

TRI-ANG T.30 AMBULANCE

TRI-ANG T.40B MONTE CARLO

Pressed-steel body, crank drive, suitable for 15" leg measurement. 7" dia. disc wheels, electric lights and streamlined horn. Finished in two-colour enamel and chrome. Length 36".

TRI-ANG T.30C TIPPER

New type pressed-steel body, 7" dia. wheels with chromed hub caps. Tipper operated by lever from driving seat. Crank drive, suitable for 15" leg measurement. High-gloss enamel finish. Length 43".

TRI-ANG T.30 AMBULANCE

New type pressed-steel body, 7" dia. wheels with chromed hub caps. Crank drive, suitable for 15" leg measurement. Detachable stretcher. Length 35".

Pedal Motors
POLICE
SQUAD CAR
TRI-ANG T.30 POLICE CAR
FIRE DE
TRI-ANG T.30 FIRE CHIEF
THUNDERBOLT
TRI-ANG T.30A THUNDERBOLT
LIGHTNING
TRI-ANG T.30B LIGHTNING
TRI-ANG T.30 POLICE CAR
TRI-ANG T.30 FIRE CHIEF
TRI-ANG T.30A THUNDERBOLT
TRI-ANG T.30B LIGHTNING

TRI-ANG TRIUMPH HERALD "B"
TRI-ANG T.45C MIAMI
TRI-ANG T.45A
TRI-ANG T.45B
Tri-ang
TRI-ANG T.45C MIAMI
TRI-ANG T.45B
TRI-ANG T.45A
TRI-ANG TRIUMPH HERALD "B"
TRI-ANG TRIUMPH HERALD "A"

Tri-ang **PEDAL MOTORS**

Regd. Trade Mark

Into the 1962 range of Pedal Motor Cars we introduce the Veteran Car. Authentically designed and beautifully finished in bright enamels it takes its place in the range of Tri-ang Pedal Motors—World renowned for their outstanding design and excellent finish; coupled with refinements which are bound to appeal to all. Opening bonnets and boots, dummy engines with oil and water tanks etc. etc.

VETERAN MOTOR CAR

36" overall pressed metal body, with adjustable crank drive. 10" front and 12" rear detachable direct spoke wheels with black cushion tyres, fitted with individual mudguards. Plastic welded seat. Accessories include 2 front lamps and rubber bulb horn. Starting handle with ratchet noise.

Page One

Tri-ang TM0200 Traveller

Tri-ang TM6806 Tufty learner's car

Tri-ang
Regd Trade Mark
Toys
take the
Tough Guy Test

Tri-ang

BEGINNER'S SERIES

1. **AUSTIN A.40** New model in modern plastic body styling. Nylon bearings. Tubular metal bumpers at front and rear. Inside leg measurement 14in.-16in. Overall length 34in.
2. **HERALD 'B'** Duotone body, chromed laced wheels and electric lights. Inside leg measurement 14in.-16in. Overall length 28in.

   HERALD 'A' (*not illustrated*) As model 'B' but in single colour and without lights. Enamelled wheels.
3. **ROYAL PRINCE** Strong steel body, 7in. disc wheels with 1in. tyres. Chromed radiator, windscreen, steering wheel and hub caps. Bulb horn. Inside leg measurement 16in.-18in. Overall length 30in.
4. **DUKE NO. 1** As Royal Prince but all enamelled finish.
5. **COMET 'A'** Suitable 'first' model with inside leg measurement 14in.-16in. Disc wheels with ½in. tyres. Overall length 26in.

Page 6

Tri-ang

1. **THE GENERAL** Strong steel body with 8in. balloon-type disc wheels. Complete with military insignia, jerri-can, spare wheel and dummy telephone. Nylon bearings. Inside leg measurement 14in.-17in. Overall length 40in.
2. **JEEP** As "General" model, less insignia and telephone. White wheels.
3. **A.A. SCOUT CAR** Brightly coloured model with electric lights. Crank drive, suitable for 14in.-17in. inside leg measurement. Nylon bearings. Overall length 37in.
4. **CANNON** For towing behind jeeps. Fires ping-pong balls. Overall length 21in.

Page 5

Tri-ang

B.M.C. SERIES

1. **WOLSELEY** Farina-styled body in duotone. Chromed radiator. Die-cast front for extra strength. Chromed laced wheels. Opening bonnet and boot. Electric lights, flashing trafficators, horn, "radio" and dummy engine with petrol, oil and water filling tanks. Adjustable seat back giving inside leg measurements of 16in.-20in. Overall length 42in.
2. **WOLSELEY 1500** As Wolseley, less flashers and petrol, oil and water tanks. Inside leg measurement 16in.-20in. Overall length 42in.
3. **MORRIS OXFORD** Farina-styled body in duotone with chromed radiator and laced wheels. Die-cast front for extra strength. Opening boot and bonnet. Electric lights and horn. Inside leg measurement 16in.-20in. Overall length 42in.
4. **AUSTIN A.50** Popular version of Farina-styled body, available in single colours. Enamelled finished radiator and wheels. Inside leg measurement 16in.-20in. Overall length 41in.

All above cars available in blue or red finish

Page 2

Tri-ang

1. **TRI-ANG KART 'B'** New model of popular Go-Kart. Strong tubular chassis. Dummy engine. Chromed laced wheels, with heavy tyres at rear. Leg measurement 17in.-19in. Overall length 44in.

   TRI-ANG KART 'A' (*not illustrated*) As Model 'B' but with enamelled wheels (all with 1in. tyres).
2. **SUPERSONIC PLANE** Streamlined model in flamboyant colour. Fitted with front and rear navigation lights. Laced wheels. Nylon bearings. Inside leg measurement 17in.-19in. Overall length 49in.
3. **ASTRONAUT** Front-drive model in flamboyant blue. Chromed handlebars. 12in. front and 7in. rear laced wheels with extra wide tyres. Adjustable sprung saddle. Nylon bearings. Safe plastic mudguard. Inside leg measurement 14in.-16in.
4. **VENUS** New model in flamboyant blue with battery-operated push-button siren. Chromed handlebars and chariot safety surround bar. Adjustable sprung saddle. Nylon bearings. Knuckle guards and streamers. Inside leg measurement 15in.-17in.
5. **SIREN** Popular model in flamboyant gold finish with siren, chariot bar and with nylon bearings. Inside leg measurement 15in.-17in.
6. **MARS** As No. 5 less siren, flamboyant red with metal saddle.

Page 8

Tri-ang

1. T.45B

2. T.45A

3. T.30 Police Car

4. T.30B Lightning

Nylon bearings and Independent suspension

5. T.30A Thunderbolt

1. **T.45B** Top seller in modern, duotone body, chromed radiator and laced wheels. Electric lights and horn. Nylon bearings. Inside leg measurement 14in.-16in. Overall length 34in.

T.45X (*not illustrated*) As T.45B, but in single colour and without lights and horn and seat pad.

2. **T.45A** As T.45X in single colour and with enamelled radiator and disc wheels.

3. **T.30 POLICE CAR** Popular model with chromed bell, radiator and hub caps. 7in. wheels with 1in. tyres. Crank drive suitable for 15in. leg measurement. Overall length 35in.

4. **T.30B LIGHTNING** Popular model with chromed radiator and hub caps. Bulb horn. 7in. wheels, with 1in. tyres. Inside leg measurement 15in. Overall length 35in.

5. **T.30A THUNDERBOLT** As T.30B but in enamelled finish and with ½in. tyres.

Tri-ang

TRACTORS

1. Ferguson Tractor

3. Farmer's Boy Tractor

NEW!

2. Harvester Tractor

Nylon bearings

4. Midget Tractor

1. **FERGUSON TRACTOR** Strong model of the popular tractor in new style blow-moulded plastic. Mounted on strong, tubular chassis with realistic 10in. front and 16in. rear wheels. Ball-bearing adjustable crank drive. Inside leg measurement 19in.-21in. Overall length 37in.

2. **HARVESTER TRACTOR** Four-wheel version of the Farmer's Boy Tractor with crank drive and strong front axle. 7½in. front and 12in. rear wheels. Overall length 28in.

3. **FARMER'S BOY TRACTOR** Light but very strong model in new styled blow-moulded plastic. Nylon bearings. Front drive action with 7½in. front and 12in. rear wheels. Inside leg measurement 14in.-16in. Overall length 28in.

4. **MIDGET TRACTOR** Metal body on strong, tubular chassis. 7in. front and 8½in. rear wheels. Nylon bearings. Inside leg measurement 13in.-15in. Overall length 33in.

Nylon bearings

Tri-ang

1. T.65 Motor

NEW!

3. T.45 Police Car

2. T.45C Miami

NEW!

4. T.45 Fire Engine

1. **T.65 MOTOR** Ever-popular model in duotone with balloon type disc wheels with rubber tyres. Electric lights and horn. Inside leg measurement 14in.-16in. Overall length 39in.

2. **T.45C MIAMI** Duotone body with opening bonnet and boot, dummy engine, electric lights and horn. Plastic steering wheel, chromed radiator and laced wheels. Inside leg measurement 14in.-16in. Overall length 34in.

T.50 MOTOR (*not illustrated*) As T65 but in single colour and without lights and horn.

3. **T.45 POLICE CAR** New model fitted with electric lights and siren. Chromed laced wheels and radiator. Inside leg measurement 15in.-17in. Overall length 34in.

4. **T.45 FIRE ENGINE** New model complete with ladders and chromed bell. Inside leg measurement 15in.-17in. Overall length 34in.

Tri-ang

1. Veteran

2. Jaguar Mk. X

Nylon bearings and Independent suspension

NEW!

3. M.G. Midget

4. Grand Prix Racer

1. **VETERAN** Replica of a car seen at the turn of the century. Adjustable crank drive, fitted with 10in. front and 12in. rear wheels. Plastic front lamps and bulb horn. Inside leg measurement 15in.-17in. Overall length 36 in.

2. **JAGUAR Mk.X** Plastic vacuum-formed body with chromed radiator, electric lights and adjustable crank drive. Nylon bearings and independent suspension on rear wheels. Inside leg measurements 14in.-16in. Overall length 48in.

3. **M.G. MIDGET** New style, plastic-formed body with M.G. radiator, electric lights and adjustable crank drive. Strong chassis with nylon bearings and independent suspension of rear wheels. Inside leg measurement 16in.-18in. Overall length 48in.

4. **GRAND PRIX RACER** Detailed pressed steel body, colours red or silver. Chromed 'exhaust', sports wheels with 2in. tyres and imitation knock-off hub caps. Safety-type steering wheel. Nylon bearings and independent suspension on rear wheels. Inside leg measurements 19in.-21in. Overall length 48in.

Tri-ang
Regd. Trade Mark
Toys manufactured at MERTON factory

☆ *Nylon bearings and Independent suspension*

TM0720 Florida (T.45B)

TM0905 Squad Car (T.30)

NEW!

TM0725 Motorway Patrol

TM0910 Lightning (T.30B)

TM0915 Thunderbolt (T.30A)

TM0720 Florida (T.45B) Modern shaped body with electric lights and horn. Chromed radiator and wheels. Inside leg measurement 14–16 in. Overall length 34 in. In Blue/White or Red/White.

TM0725 Motorway Patrol Chromed radiator and bell. Enamelled laced wheels. Inside leg measurement 14–16 in. Overall length 34 in.

TM0730 Bermuda (T.45A) In single colour with enamelled radiator and *now with laced wheels.* Inside leg measurement 14–16 in. Overall length 34 in. Available in Blue or Red.

TM0905 Squad Car (T.30) Chromed bell and radiator. 7 in. disc wheels. Inside leg measurement 15–17 in. Overall length 35 in.

TM0910 Lightning (T.30B) Chromed radiator, bulb horn. 7 in. wheels with 1 in. tyres. Inside leg measurement 15–17 in. Overall length 35 in. Available in Blue.

TM0915 Thunderbolt (T.30A) As 'Lightning' but with enamelled radiator and fitted ½ in. tyres. Available in Red.

Page 4

Tri-ang TM0205 Clubman

Tri-ang TM1515 Police Jeep

Tri-ang Toys manufactured at MERTON factory

☆ *Nylon bearings and Independent suspension*

TM0120 Grand Prix Racer

TM0125 Veteran

NEW!

TM0130 'E' Type Jaguar

NEW!

TM0135 Triumph TR.4

TM0120 Grand Prix Racer Steel body. Available in Red or Silver finish. Sports type wheels with 2 in. tyres. Inside leg measurement 19–21 in. Overall length 48 in.

TM0125 Veteran Strong body with 10 in. front and 12 in. rear direct spoke wheels. Plastic front lamps and bulb horn. Adjustable crank drive. Inside leg measurement 15–17 in. Overall length 36 in.

TM0130 'E' Type Jaguar Shaped steel body with polychromatic finish. Bucket seat. Chromed laced wheels. Inside leg measurement 16–18 in. Overall length 41 in.

TM0135 Triumph TR.4 Injection moulded body in strong plastic. Chromed laced wheels and tubular bumpers. Plated radiator. Inside leg measurement 16–18 in. Overall length 46 in. Available in yellow or red.

Page 2

Tri-ang Toys manufactured at MERTON factory

NEW!

TM0105 Rolls Royce

TM0110 M.G. Midget

☆ *Nylon bearings and Independent suspension*

TM0115 Jaguar Mk. X

TM0105 Rolls Royce Luxury model in moulded plastic. 6v. printed circuit with independent switches for headlamps and side and tail lights combined. Chromed rim embellishers. New type windscreen and dummy hood. Thick rubber tyres. Available in Grey, Maroon or Black. Inside leg measurement 16–18 in. Overall length 51 in.

TM0110 M.G. Midget Plastic formed body with electric lights. Chromed sports wheels. Inside leg measurement 16–18 in. Overall length 48 in. Available in Blue or Red.

TM0115 Jaguar Mk. X Plastic formed body with electric lights and adjustable crank drive. Inside leg measurement 14–16 in. Overall length 48 in. Available in Blue or Red.

Page 1

Lines Bros. Ltd., Tri-ang Works, Morden Road, Merton, London, S.W.19

TM1105 Austin A.40

TM1110 Royal Prince

TM1505 General

TM1115 Duke

TM1510 Jeep

TM1110 Royal Prince Strong steel body with chromed radiator, enamelled laced wheels, 1 in. tyres. Inside leg measurement 16–18 in. Overall length 30 in. Available in Blue.

TM1115 Duke As 'Royal Prince' but with enamelled finish. (¾ in. tyres). Available in Red.

TM1105 Austin A.40 Plastic body with tubular bumpers front and rear. Nylon bearings. Inside leg measurement 14–16 in. Overall length 34 in.

TM1505 General Strong steel body. Balloon type disc wheels. Military insignia, jerri-can, spare wheel and dummy telephone. Nylon bearings. Inside leg measurement 14–17 in. Overall length 40 in.

TM1510 Jeep As 'General' model less insignia, telephone and spare wheel.

Page 5

Lines Bros. Ltd., Tri-ang Works, Morden Road, Merton, London, S.W.19

☆ Nylon bearings and Independent suspension

TM0505 Wolseley

TM0510 Morris Isis

NEW!

TM0515 Austin A.50

TM0705 Miami (T.45C)

Above cars are available in Blue or Red with duotone finish where shown.

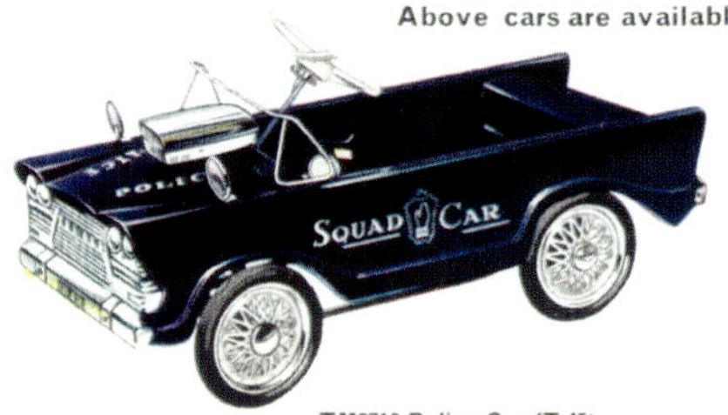

TM0710 Police Car (T.45)

TM0505 Wolseley Farina styled body with strong die-cast front. Opening bonnet and boot and including electric lights, flashing indicators, electric horn, 'radio' and dummy engine with petrol, oil and water filling tanks. Adjustable seat back giving inside leg measurement 16–20 in. Overall length 42 in.

TM0515 Austin A.50 In single colour with enamel finished radiator and wheels. Inside leg measurement 16–20 in. Overall length 41 in.

TM0710 Police Car (T.45) Complete with electric lights and siren. Chromed radiator and wheels. Inside leg measurement 15–17 in. Overall length 34 in.

TM0510 Morris Isis Farina styled body with opening bonnet and boot. Complete with electric lights, electric horn and dummy engine. Inside leg measurement 16–20 in. Overall length 42 in.

TM0705 Miami (T.45C) With opening boot and bonnet, dummy engine, electric lights and horn. Chromed radiator and wheels. Inside leg measurement 14–16 in. Overall length 34 in.

TM0715 Fire Engine (T.45) Including ladders, chromed bell and with enamelled radiator. *Now with laced wheels.* Inside leg measurement 15-17 in. Overall length 34 in.

Page 3

Tri-ang Tractor/dump truck (ref. no. unknown)

Tri-ang TM0107 Safari Overlander

Tri-ang Toys manufactured at MERTON factor

TM1105 Austin A.40 Plastic body with tubular bumpers front and rear. Nylon bearings. Inside leg measurement 14-16 in. Overall length 35 in.

TM1505 General Jeep Strong steel body. Balloon type disc wheels. Military insignia, jerri-can, spare wheel and dummy telephone. Nylon bearings. Inside leg measurement 14-17 in. Overall length 40 in.

TM1515 Police Jeep Strong steel body in police Blue with flashing electric light. Balloon type disc wheels. Inside leg measurement 14-17 in. Overall length 40 in.

TM1510 Jeep As 'General' model less insignia, telephone and spare wheel.

Page 6

Tri-ang Toys manufactured at MERTON factory

☆ Nylon bearings and Independent suspension

NEW!

TM0120 Grand Prix Racer

TM0121 Junior Grand Prix Racer

NEW!

TM0130 'E' Type Jaguar

TM0205 Tri-ang '2000'

TM0120 Grand Prix Racer Steel body. Available in Red or polychromatic Blue. Sports type wheels with 2 in. tyres. Inside leg measurement 19-21 in. Overall length 48 in.

TM0130 'E' Type Jaguar Shaped steel body with polychromatic Blue or Green finish. Bucket seat. Chromed laced wheels. Inside leg measurement 16–18 in. Overall length 41 in.

TM0136 Super Triumph TR.4 *(not illustrated)* Injection moulded body in strong plastic. Chromed laced wheels and tubular bumpers. Plated radiator. Electric lights. Inside leg measurement 16-18 in. Overall length 46 in. Available in Yellow or Red.

TM0135 Triumph TR.4 *(not illustrated)* As 'Super' but minus electric lights and in all-enamel finish.

TM0137 Electric Triumph TR.4 *(not illustrated)* As 'Super' but electrically driven by a leak-proof rechargeable battery. Hand-operated control on dashboard for forward and reverse. Maximum speed 3 m.p.h.

TM0150 Chargemobile *(not illustrated)* For charging battery of Electric TR.4 direct from mains supply.

TM0121 Junior Grand Prix Racer As Grand Prix Racer but minus air-intake and exhaust. Enamelled laced wheels with extra-thick tyres.

TM0205 Tri-ang '2000' Pressed steel body of advanced design in duo-tone finish with chromed radiator and laced wheels. Combination electric light/horn unit. Inside leg measurement 15–19 in. Overall length 40 in.

TM0210 Tri-ang '1500' *(not illustrated)* As '2000' but in single colour and minus light/horn unit. Transfers on sides.

TM0215 Tri-ang '1000' *(not illustrated)* As '1500' but with silvered wheels, radiator and back and minus windscreen.

Page 2

Tri-ang Toys manufactured at MERTON factory

TM0106 Rolls Royce

☆ Nylon bearings and Independent suspension

TM0111 M.G. Midget

M.G. Midget and Jaguar MK X now fitted with electric light horn unit.

TM0116 Jaguar Mk. X

TM0106 Rolls Royce Luxury model in moulded plastic. 6v. printed circuit with switch for headlamps and independent switch for side and tail lights. Electric horn. Chromed rim embellishers. New type windscreen and dummy hood. Thick rubber tyres. Available in Grey, Maroon or Black. Inside leg measurement 16–18 in. Overall length 51 in.

TM0111 M.G. Midget Plastic formed body now fitted with electric light/horn unit on bonnet. Chromed sports wheels. Inside leg measurement 16–18 in. Overall length 48 in. Available in Blue or Red.

TM0116 Jaguar Mk. X Plastic formed body now fitted with electric light/horn unit on bonnet. Inside leg measurement 14–16 in. Overall length 48 in. Available in Blue or Red.

Page 1

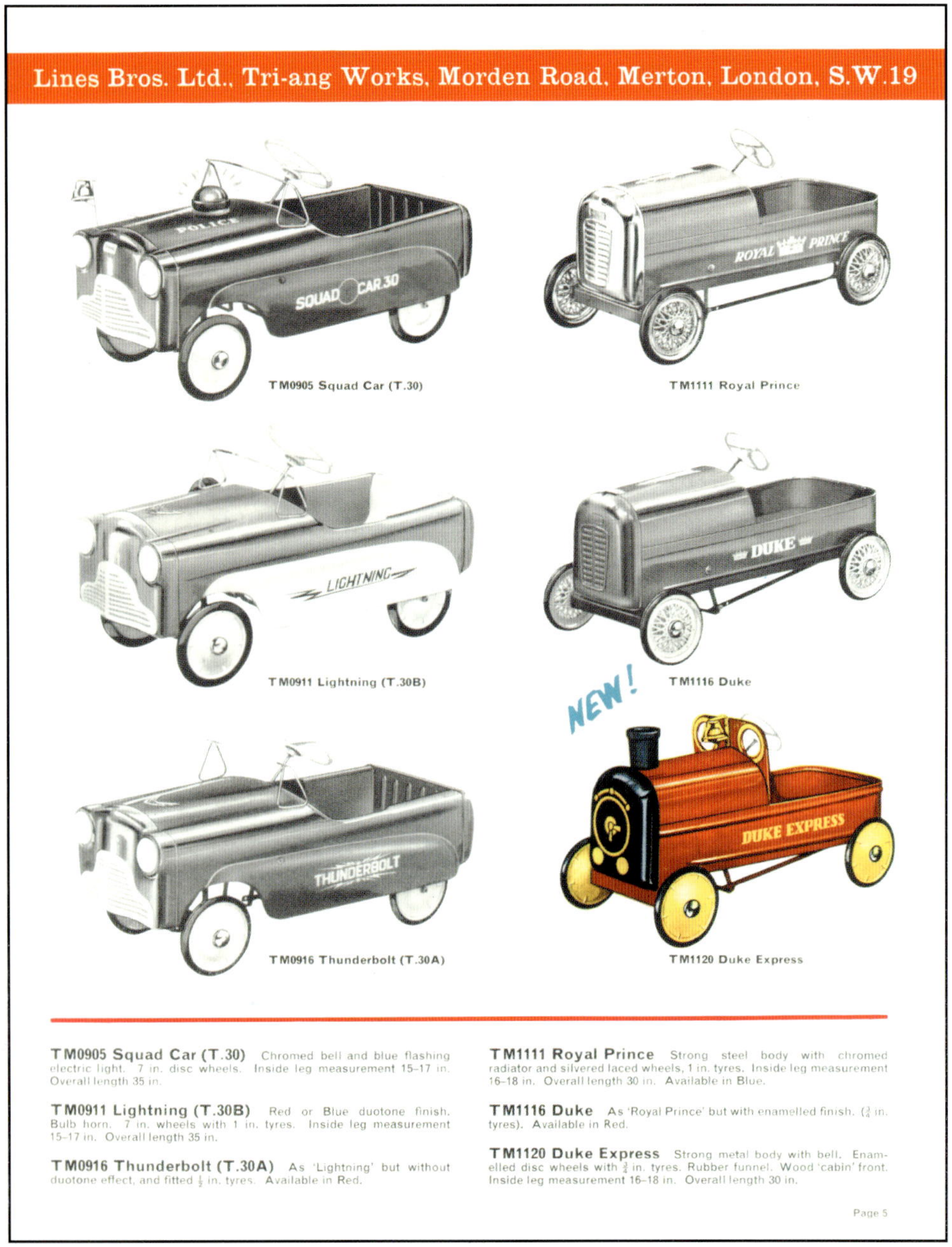

Lines Bros. Ltd., Tri-ang Works, Morden Road, Merton, London, S.W.19

TM0905 Squad Car (T.30)

TM1111 Royal Prince

TM0911 Lightning (T.30B)

NEW!

TM1116 Duke

TM0916 Thunderbolt (T.30A)

TM1120 Duke Express

TM0905 Squad Car (T.30) Chromed bell and blue flashing electric light. 7 in. disc wheels. Inside leg measurement 15–17 in. Overall length 35 in.

TM0911 Lightning (T.30B) Red or Blue duotone finish. Bulb horn. 7 in. wheels with 1 in. tyres. Inside leg measurement 15–17 in. Overall length 35 in.

TM0916 Thunderbolt (T.30A) As 'Lightning' but without duotone effect, and fitted ½ in. tyres. Available in Red.

TM1111 Royal Prince Strong steel body with chromed radiator and silvered laced wheels, 1 in. tyres. Inside leg measurement 16–18 in. Overall length 30 in. Available in Blue.

TM1116 Duke As 'Royal Prince' but with enamelled finish. (¾ in. tyres). Available in Red.

TM1120 Duke Express Strong metal body with bell. Enamelled disc wheels with ¾ in. tyres. Rubber funnel. Wood 'cabin' front. Inside leg measurement 16–18 in. Overall length 30 in.

Page 5

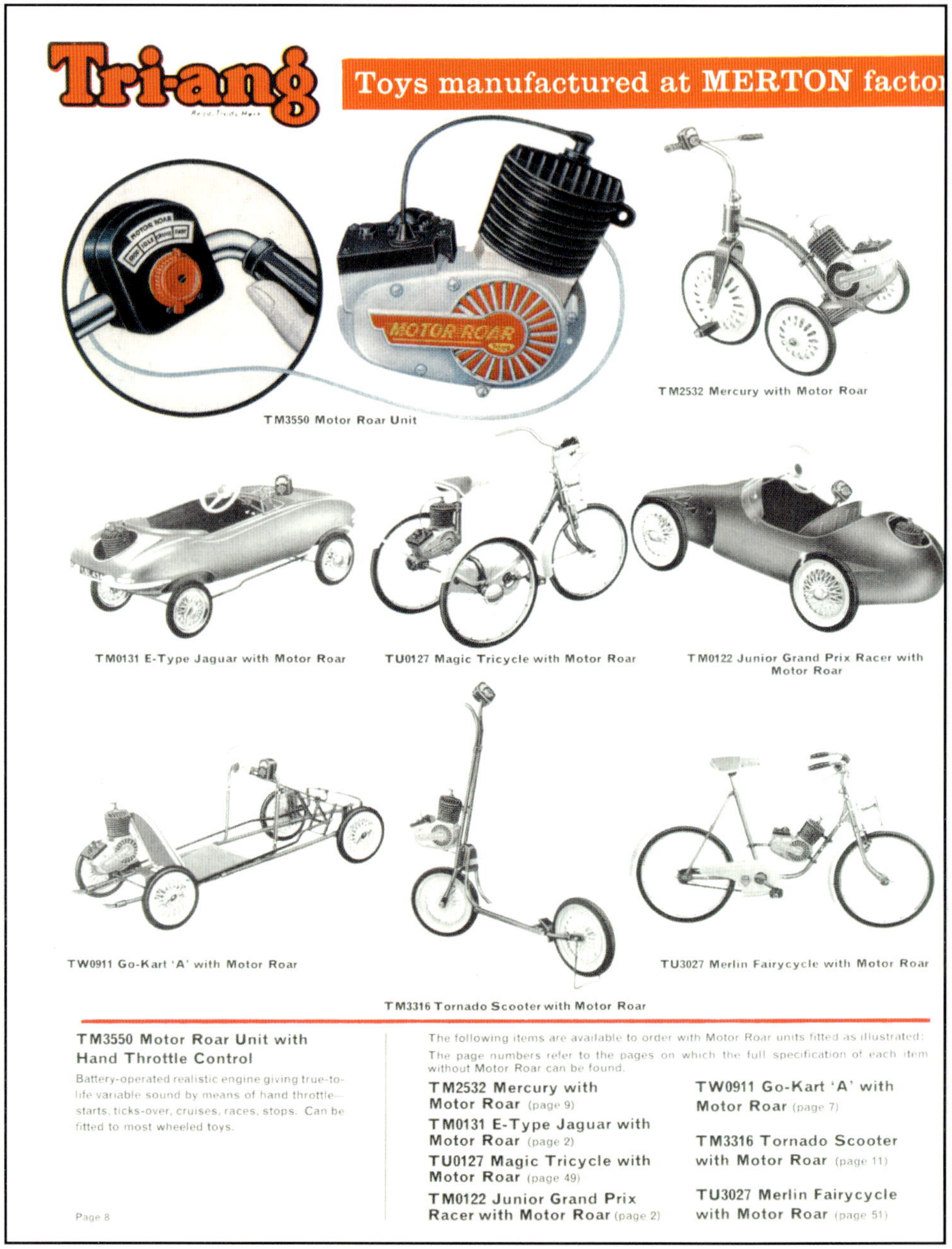

Tri-ang

Toys manufactured at MERTON factor

TM3550 Motor Roar Unit

TM2532 Mercury with Motor Roar

TM0131 E-Type Jaguar with Motor Roar

TU0127 Magic Tricycle with Motor Roar

TM0122 Junior Grand Prix Racer with Motor Roar

TW0911 Go-Kart 'A' with Motor Roar

TU3027 Merlin Fairycycle with Motor Roar

TM3316 Tornado Scooter with Motor Roar

TM3550 Motor Roar Unit with Hand Throttle Control

Battery-operated realistic engine giving true-to-life variable sound by means of hand throttle—starts, ticks-over, cruises, races, stops. Can be fitted to most wheeled toys.

The following items are available to order with Motor Roar units fitted as illustrated:

The page numbers refer to the pages on which the full specification of each item without Motor Roar can be found.

TM2532 Mercury with Motor Roar (page 9)

TM0131 E-Type Jaguar with Motor Roar (page 2)

TU0127 Magic Tricycle with Motor Roar (page 49)

TM0122 Junior Grand Prix Racer with Motor Roar (page 2)

TW0911 Go-Kart 'A' with Motor Roar (page 7)

TM3316 Tornado Scooter with Motor Roar (page 11)

TU3027 Merlin Fairycycle with Motor Roar (page 51)

Page 8

Tri-ang

Toys manufactured at MERTON factory

☆ *Nylon bearings and Independent suspension*

TM0720 Florida (T.45B)

TM0751 Bahama

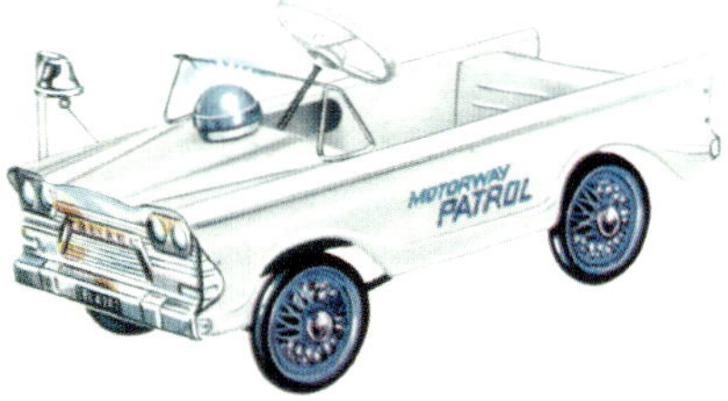

TM0726 Motorway Patrol

NEW!

TM0735 Pick-up Truck

TM0731 Bermuda (T.45A)

TM0720 Florida (T.45B) Modern shaped body with electric lights and horn. Chromed radiator and wheels. Inside leg measurement 14–16 in. Overall length 34 in. In Blue/White or Red/White.

TM0726 Motorway Patrol Chromed radiator and bell. Blue flashing electric light. Enamelled laced wheels. Inside leg measurement 14–16 in. Overall length 34 in.

TM0731 Bermuda (T.45A) In single colour with enamelled radiator and laced wheels. Inside leg measurement 14–16 in. Overall length 34 in. Available in Blue or Red.

TM0751 Bahama In single colour with chromed radiator and laced wheels. Electric horn. Inside leg measurement 14–16 in. Overall length 34 in.

TM0735 Pick-up Truck White enamelled radiator and laced wheels. Flat platform behind seat for carrying loads. Inside leg measurement 14–16 in. Overall length 34 in.

Page 4

Lines Bros. Ltd., Tri-ang Works, Morden Road, Merton, London, S.W.19

TM2005 Ferguson Tractor

TM2010 Harvester Tractor

NEW!

TM2100 Dumper Truck

TW0905 Kart 'B'

TW6035 Trailer

TW0910 Kart 'A'

TM2005 Ferguson Tractor Strong model in moulded plastic mounted on tough tubular chassis with realistic 10 in. front and 16 in. rear wheels. Ball-bearing adjustable crank drive. Inside leg measurement 19–21 in. Overall length 37 in.

TM2100 Dumper Truck Strong tubular frame with metal pan fitted with tipping handle. 12 in. wheels. Inside leg measurement 21–23 in. Overall length 44 in.

TM2015 Farmer's Boy Tractor *(not illustrated)* Similar to Harvester Tractor but with front wheel pedal drive. Nylon bearings. 7½ in. front and 12 in. rear wheels. Inside leg measurement 14-16 in. Overall length 28 in.

TW6035 Trailer Suitable for towing behind tractors.

TM2010 Harvester Tractor Light but strong model in moulded plastic with crank drive. 7½ in. front and 12 in. rear wheels. Inside leg measurement 15–17 in. Overall length 28 in.

TW0905 Kart 'B' Strong tubular chassis with all-black dummy engine and chromed laced wheels with heavy rear tyres. Adjustable seat giving leg measurement 17–21 in. Overall length 44 in.

TW0910 Kart 'A' As Kart 'B' but with enamelled sports wheels (all 1 in. tyres), shaped wood seat, and less engine. Inside leg measurement 17–21 in. Overall length 42 in.

Page 7

Lines Bros. Ltd., Tri-ang Works, Morden Road, Merton, London, S.W.19

☆ Nylon bearings and Independent suspension

TM0506 Wolseley

TM0510 Morris Isis

TM0516 Austin A.50

TM0705 Miami (T.45C)

Above cars are available in single colours or with duotone finish where shown.

TM0710 Police Car (T.45)

TM0715 Fire Engine (T.45)

TM0506 Wolseley Farina styled body with strong die-cast front. Opening bonnet and boot and including electric lights, flashing indicators, electric horn, 'radio' and dummy engine. Adjustable seat back giving inside leg measurement 16–20 in. Overall length 42 in.

TM0516 Austin A.50 In single colour with silvered radiator and wheels. Electric light/horn unit. Inside leg measurement 16–20 in. Overall length 41 in.

TM0710 Police Car (T.45) Complete with electric lights and siren. Chromed radiator and wheels. Inside leg measurement 15–17 in. Overall length 34 in.

TM0510 Morris Isis Farina styled body with opening bonnet and boot. Complete with electric lights, electric horn and dummy engine. Inside leg measurement 16–20 in. Overall length 42 in.

TM0705 Miami (T.45C) With opening boot and bonnet, dummy engine, electric lights and horn. Chromed radiator and wheels. Inside leg measurement 14–16 in. Overall length 34 in.

TM0715 Fire Engine (T.45) Including ladders, chromed bell and with enamelled radiator. *Now with laced wheels.* Inside leg measurement 15–17 in. Overall length 34 in.

Page 3

Tri-ang Pedal boat
(ref. no. unknown)

Tri-ang Armoured car sit
and ride (ref. no. unknown)

Tri-ang small dumper (ref. no. unknown)

Tri-ang TM1120 Duke Express

TM0132 'BLUEBIRD' PEDAL BOAT Exciting new model in strong plastic with timber decking fore and aft. Fitted with swivelling searchlight with press-button for morse signalling; mooring rope, wrap-round windscreen, enamel laced wheels. Inside leg measurement 16 in. Overall length 41 in.

TM0106 ROLLS ROYCE Luxury model in moulded plastic. 6v. printed circuit with switch for headlamps and independent switch for side and tail lights. Electric horn. Chromed rim embellishers. New type windscreen and dummy hood. Thick rubber tyres. Inside leg measurement 16–18 in. Overall length 51 in.

TM0107 SAFARI OVERLANDER Pressed steel Jeep body with trailer. Shovel and cap-firing rifle strapped to sides and real camera fitted over windscreen. Jerri-can. Ridge tent in its own canvas sack is carried in trailer. Inside leg measurement 14–17 in. Overall length of Jeep and Trailer 64 in.

Page 1

Tri-ang TOYS MERTON

☆ Nylon bearings and Independent suspension

NEW!

TM0120 GRAND PRIX RACER

TM0109 CORSAIR

TM0130 'E' TYPE JAGUAR

TM0111 M.G. MIDGET

TM0109 CORSAIR Plastic formed body with chromed laced wheels with large hub caps. Chromed strip along body and chromed wrap-round bumpers. Detailed printed fascia panel. Inside leg measurement 18 in. Overall length 47 in.

TM0111 M.G. MIDGET Plastic formed body fitted with electric light/horn unit on bonnet. Chromed sports wheels. Inside leg measurement 16–18 in. Overall length 48 in. Available in Blue or Red.

TM0120 GRAND PRIX RACER Steel body. Available in Red or Polychromatic Blue. Sports type wheels with 2 in. tyres. Inside leg measurement 19–21 in. Overall length 46 in.

TM0121 JUNIOR GRAND PRIX RACER *(not illustrated)* As Grand Prix Racer but minus air-intake and exhaust. Enamelled laced wheels with extra-thick tyres.

TM0130 'E' TYPE JAGUAR Shaped steel body with polychromatic Blue or Green finish. Bucket seat. Chromed laced wheels. Inside leg measurement 16–18 in. Overall length 41 in.

Page 2

Tri-ang TOYS MERTON

NEW!

TM0720 FLORIDA

TM0070 FIRE SERVICE TRUCK

☆ Nylon bearings

TM0726 MOTORWAY PATROL

TM0705 MIAMI

TM0729 POLICE SQUAD CAR

NEW!

TM0712 DAIRYMAN

TM0731 BERMUDA

TM0700 FIRE SERVICE TRUCK Pressed steel body with Red enamel finish and white silk screening. Fitted with 30 ft. of plastic hose pipe with jet nozzle and water tap fittings. Rotating hose wheel. Siren and Blue flashing light. Inside leg measurement 16–18 in. Overall length 34 in.

TM0705 MIAMI With opening boot and bonnet, dummy engine, flashing amber indicators, and horn. Chromed radiator and wheels.

TM0712 DAIRYMAN Blue enamelled laced wheels and Blue sprayed radiator. Platform at rear carries 4 plastic milk churns.

TM0720 FLORIDA Modern shaped body now fitted with electric light/horn unit on bonnet. Chromed radiator and wheels.

TM0726 MOTORWAY PATROL Chromed radiator and bell. Blue flashing electric light. Enamelled laced wheels.

TM0729 POLICE SQUAD CAR All Navy body with White silk screening on sides. Chromed bell.

TM0731 BERMUDA In single colour with White disc wheels.

Page 4

NEW!

TM1100 INDIAN CHIEF

TM1120 DUKE EXPRESS

TM1111 ROYAL PRINCE

TM1505 GENERAL JEEP

TM1116 DUKE

TM1515 POLICE JEEP

TM1100 INDIAN CHIEF Pedal-driven engine steered by cross-bar over 'cab'. Front in moulded plastic with shaped rubber funnel. Chromed bell. Inside leg measurement 15–17 in. Overall length 31 in.

TM1111 ROYAL PRINCE Strong steel body with chromed radiator and silvered laced wheels, ¾ in. tyres. Inside leg measurement 16–18 in. Overall length 30 in.

TM1116 DUKE As 'Royal Prince' but with enamelled finish. (¾ in. tyres).

TM1120 DUKE EXPRESS Strong metal body with bell. Enamelled disc wheels with ¾ in. tyres. Rubber funnel. Wood 'cabin' front. Inside leg measurement 16–18 in. Overall length 30 in.

TM1505 GENERAL JEEP Strong steel body. Balloon type disc wheels. Military insignia, jerri-can, spare wheel and dummy telephone. Nylon bearings. Inside leg measurement 17–19 in. Overall length 38 in.

TM1510 JEEP *(not illustrated)* As 'General' model less insignia, telephone and spare wheel.

TM1515 POLICE JEEP Strong steel body in police Blue with flashing electric light. Balloon type disc wheels.

Page 5

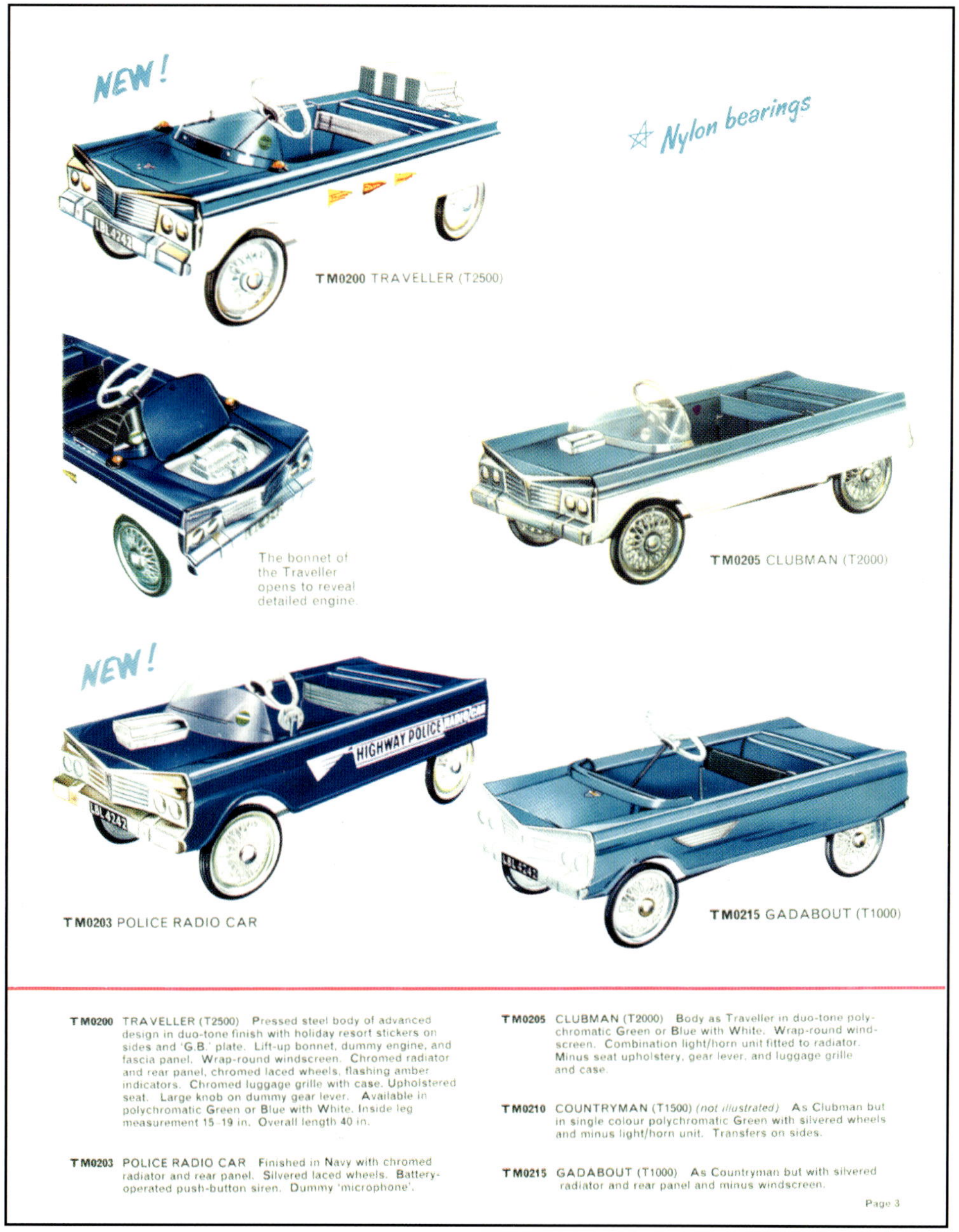

NEW!

☆ Nylon bearings

TM0200 TRAVELLER (T2500)

The bonnet of the Traveller opens to reveal detailed engine.

TM0205 CLUBMAN (T2000)

NEW!

TM0203 POLICE RADIO CAR

TM0215 GADABOUT (T1000)

TM0200 TRAVELLER (T2500) Pressed steel body of advanced design in duo-tone finish with holiday resort stickers on sides and 'G.B.' plate. Lift-up bonnet, dummy engine, and fascia panel. Wrap-round windscreen. Chromed radiator and rear panel, chromed laced wheels, flashing amber indicators. Chromed luggage grille with case. Upholstered seat. Large knob on dummy gear lever. Available in polychromatic Green or Blue with White. Inside leg measurement 15–19 in. Overall length 40 in.

TM0203 POLICE RADIO CAR Finished in Navy with chromed radiator and rear panel. Silvered laced wheels. Battery-operated push-button siren. Dummy 'microphone'.

TM0205 CLUBMAN (T2000) Body as Traveller in duo-tone polychromatic Green or Blue with White. Wrap-round windscreen. Combination light/horn unit fitted to radiator. Minus seat upholstery, gear lever, and luggage grille and case.

TM0210 COUNTRYMAN (T1500) *(not illustrated)* As Clubman but in single colour polychromatic Green with silvered wheels and minus light/horn unit. Transfers on sides.

TM0215 GADABOUT (T1000) As Countryman but with silvered radiator and rear panel and minus windscreen.

Page 3

Tri-ang TM0215 Gadabout

Tri-ang TW3145 Formula 2 racing car

TM 0109
Corsair. A great model of a real man's car! Tough and racily handsome body of formed plastic with gleaming chrome everywhere – on the laced wheels, large hub caps, body trim and wrap-around bumpers. Authentically printed fascia panel shows all the instruments. New-type oval white steering wheel.
ILM: 18 ins. OL. 47 ins.

* New-type steering wheels
* Independent suspension
* Nylon bearings

TM 0120
Grand Prix Racer. A fabulous racer for get-away boys. Strong steel body in a choice of red or polychromatic green finish. The real sports type wheels are fitted with 2 in. tyres. The new-type oval white steering wheel is fitted. The driver is protected by a wrap-around windscreen. Chromed air-intake on bonnet, chromed exhaust and "knock-off" hub caps are details an up-and-coming driver will notice. This tough, handsome toy is terrific value.
ILM: 19 ins-21 ins. OL: 46 ins.

TM 0121
Junior Grand Prix Racer (not illustrated). As Grand Prix Racer but minus air-intake on bonnet and exhaust. Large enamelled laced wheels with thick tyres.

TM 0106
Rolls Royce. A fabulous model of the world's finest car – correct in every detail! Moulded from luxurious looking, long-lasting plastic. A real winner. Has 6V printed circuit with switch for headlamps and independent switch for side/tail lamps. Electric horn. Thick rubber tyres and a new type of windscreen and dummy hood. Rim embellishers in sparkling chrome.
ILM: 16 ins-18 ins. OL: 51 ins.

Tri-ang TOYS LEAD

2

M

TM 0130
'E' Type Jaguar. The famous model that is every boy's joy. The robust shaped steel body is finished in either bright red or blazing white. Bucket type seat and chromed laced wheels with "knock-off" hub caps. ILM: 16 ins.-18 ins. OL: 41 ins.

TW 0900
Junior Kart. Just watch it go! This robust kart has a strong tubular frame and is fitted with disc wheels with chromed hub caps. Lacquered plywood seat is adjustable. ILM: 13 ins.-15 ins. OL 31 ins.

TW 0905
Kart 'B'. This has real Go-Kart ruggedness and appearance. Strong tubular chassis, chromed laced wheels with heavy-rear tyres. Black dummy engine at rear. Adjustable seat giving ILM: 17 ins.-21 ins. OL: 44 ins.

TW 0908 NEW 1967 MODEL
Kart 'C'. Chain-drive generates real power for extra speed. Strong tubular chassis with chromed laced wheels. Adjustable padded seat giving ILM: 16 ins.-18 ins. OL: 43 ins.

* Nylon Bearings

THE WAY

M

Tri-ang
REGD TRADE MARK
TOYS

Rule the waves

TM 0134 NEW 1967 MODEL
"Sea Spray" Pedal Boat. Nobody could resist this exciting robust boat on wheels – it's built to last and stand plenty of rough treatment and still look bright and handsome. The plastic body with real timber decking and bow and stern rails gives that authentic boat look and there's a wrap-round windscreen, mooring rope, chromed laced wheels with new chromed hub caps and a swivel-searchlight with morse signalling button. New-type oval chromed steering wheel gives a final exciting touch to this superb pedal boat. ILM: 16 ins. OL: 41 ins.

TM 0132
"Bluebird" Pedal Boat. Similar to "Sea Spray" but minus bow and stern rails and with enamelled laced wheels and white oval steering wheel.

* Nylon Bearings

1

Tri-ang
REGD TRADE MARK
TOYS

TM 0136 NEW 1967 MODEL
Moon Rocket. The flights of fancy this magnificent machine will conjure up for its astronaut. strong moulded plastic body with detachable nose cone. Exciting array of instruments and push-pull steering columns.
ILM: 16 ins.-18 ins. OL: 41 ins.

TM 0200
Traveller. In polychromatic green with chromed laced wheels with new chromed hub caps. New-type oval chromed steering wheel. The bonnet lifts up to reveal a detailed dummy engine! There's an authentic fascia panel, wrap-round windscreen, upholstered seat, flashing amber indicators and chromed radiator and gear-lever. Chromed luggage grille – has a fitted case.

TM 0205
Clubman Bodywork as Traveller but in polychromatic blue and minus seat upholstery, gear lever and luggage grille with case. Wrap-round windscreen. Combination light horn unit fitted to radiator.

TM 0215
Gadabout. Body in red with white radiator/rear panel and minus windscreen. New-type oval white steering wheel.

Note: All cars on this page have ILM 15in.-19in. OL 40 ins

- ✻ New body decor
- ✻ Nylon bearings
- ✻ Independent suspension
- ✻ New-type steering wheels

4

This stylish group of top quality toys has been built around a particularly modern body design, with styling way ahead of its time. They're colourful, tough and guaranteed to please their owners. All are fitted with the new-type oval white steering wheel. Note: ILM and OL are common to the range.
ILM: 16 ins.-18 ins. OL: 34 ins.

TM 0705
Miami. Smart and bright, with bonnet and boot that open – and under the bonnet a 'dummy' engine. Chromed radiator and wheels. Horn, flashing amber indicators and new chromed hub caps complete a very special pedal car.

TM 0726
Motorway Patrol. King of the highway! Real light flashes and bell rings. Bell and radiator in gleaming chrome. Enamelled laced wheels. A top quality toy – something special

TM 0700
Fire Service Truck. Striking red enamelled body of strong pressed steel with white screened sides. Enamelled laced wheel chromed hub caps. The hose wheel rotates and has 30 ft. of plastic hose with jet nozzle and tap fittings. Siren and blue flashing light that work!

5

Tri-ang TOYS

TM 0729
Police Squad Car. Everything gives way to this handsome car in smart navy blue with white screened sides. Real police type bell with chromed finish.

TM 0720
Florida. Handsome car fitted with electric light horn unit on the bonnet. Chromed radiator and wheels.

TM 0731
Bermuda. Racy roadster, finished in a bright, durable colour with smart disc wheels.

TM 1120
Duke Express. Attractive pedal train, with tough metal body wooden cabin front and enamelled disc wheels fitted with ¾ ins. tyres. Rubber funnel and chromed bell. ILM: 16 ins.-18 ins. OL. 30 ins.

TM 1111
Royal Prince. Racy little pedal car with strong steel body, chromed radiator and laced wheels. Sturdy ¾ in. tyres. ILM: 16 ins.-18 ins. OL. 30 ins.

TM 1116
Duke. As the 'Royal Prince' but with an enamelled radiator and silvered wheels.

M

TM 1505
General Jeep. Looks tough – and is! Body of strong steel, disc wheels, nylon bearings. Military insignia, jerri-can, spare wheel and dummy 'phone give a really business-like appearance. New-type oval white steering wheel.

TM 1520 NEW 1967 MODEL
Armoured Car. Comes out of 'Tight spots' still looking very handsome. Mounted with machine gun and painted with authentic army transfers. ILM: 17 ins.-19 ins. OL: 35 ins.

TM 1525 NEW 1967 MODEL
Jungle Wagon and Truck. A big-game hunter will give an eye-tooth for this! Fitted at front with towing rope in case it gets bogged down and the trailer has a jerri-can with an emergency 'petrol' supply.

TM 1535
Jungle Camp Set. (Available separately) To add the final touch to the Jungle Wagon, the set comprises a ridge tent which comes in its own draw-string bag, cap-firing rifle, metal spade and a camera that really does take photographs.

7

Tri-ang Tricycle (ref. no. unknown)

Tri-ang TM2008 Tractor shovel

Tri-ang
British Engineered with Built-in Strength
TRI-ANG TOYS LIMITED – THE BIG TOY MAKERS

GO-KARTS
Go-go-go for big sales and happy customers—with these fabulous Tri-ang Go-Karts!
TW 3115 KART 'C'
Tough tubular steel chassis, adjustable padded seat, fast-stopping hand-operated brake on both rear wheels—and a chain drive for go-go-go speed power; and with a low centre of gravity, it's as safe as it's exciting on the corners! Wheels: 7½ in. front, 10 in. rear, all four with chromed laced hubs and red 'knock-off' hub-caps.
ILM: 16-18 in. (40,6cm.-45,7cm.)
OL: 43 in. (109,2cm).
TW 3110 KART 'B'
'Second cousin' to the Kart 'C', but with adjustable pedal drive and 9 in. rear wheels.
TW 3015 JUNIOR KART
Just watch this one go! A safe, robust kart of tubular steel with adjustable bucket seat, and pedal drive. 6½ in. wheels with disc hubs and 1½ in. hubcaps.
ILM: 12-14 in. (30,4cm.-35,5cm.) OL: 31 in. (78,7cm.)
Page 9

TW 3148 NEW FOR THE 70's

HOVERCRAFT ELECTRIC

The electric battery-powered 'Sit 'n' Ride' toy every boy will want. In smooth strong plastic, with stop-go foot-operated button. Single front-wheel steering provides fast about-turns in dogem car fashion. The battery is non-spill and re-chargeable. Seat height 8¼ in.
OL: 23 in. (58,4cm.)

The seat removed to show the non-spillable battery which is rechargeable through a mains transformer, available as an extra.

NEW FOR THE 70's **TW 3170**

CARNABY

Flower-power at its most powerful! Pulsating orange moulded body with contrasting bonnet and boot lid in Soho blue. Duotone disc hubs and fully upholstered seat in "mod" flower p.v.c. completes a car guaranteed to stop the traffic!
ILM: 16-18 in. (40,6cm - 45,6cm) OL: 38 in. (96,5cm)

TW 3150

GT SPORTS

Everything opens on this eye-opening model—doors, bonnet—and the boot! It's built to last in rugged, brilliant red polypropylene on a strong steel chassis. Big value features include: chromed tubular front/rear bumpers, plated radiator and headlamps, black steering wheel, fully upholstered seat, detailed dummy engine and printed dash with a full set of 'instruments'.
It's a knockout!
ILM: 16-18 in. (40,6cm.-45,7cm). OL: 38 in. (96,5cm.)

Page 5

TM 0145 NEW FOR THE 70's

MILITARY JEEP

Ideal transport for front-line inspection. Jeep-style body in pressed steel with tough moulded wheels and a fold-flat windscreen and moulded seat. Four-star general's flag is mounted on the moulded radiator.
ILM: 17 in.-19 in. (43,1cm.-48,2cm.)
OL: 38 in. (96,5cm.)

TW 3146

FORMULA 4 (ELECTRIC MODEL)

Great news! This fabulous big-selling racer now available, as the basic Formula 2, but with re-chargeable, battery-operated electric motor drive—forward or reverse at the flick of a switch! Other details as Formula 2. (Battery re-charging units available as an extra.)

TW 3145

FORMULA 2 RACING CAR

The most exciting toy racer available, in true Grand Prix style. The tough injection-moulded body in eye-catching red embellished with yellow front, wheels and detailed plated 'engine' at rear. The faithful styling includes a 'scoop' air intake and rear-mounted twin exhausts and port inlets. The wheels (7⅞ in. front, 8½ in. rear) have strong moulded hubs and are fitted with extra-wide tyres. A real winner!
ILM: 20-22 in. (50,8cm.-55,8cm.)
OL: 48 in. (121,9cm.)

Page 4

TW 3155
SPRITE DE LUXE

Another big seller! Junior Grand Prix drivers will revel in the sleek, low-slung body of bright red moulded polypropylene, with realistic transfers, four rear-mounted, chrome-plated exhausts, $7\frac{1}{2}$ in. wheels with plated moulded hubs, and twin rear-view mirrors.
ILM: 15-16 in. (38,1cm-40,6cm)
OL: 34 in. (86,3cm)

TW 3162
SPRITE

Bound to attract attention on the starting grid is this blue Sprite with its 9 in. blow-moulded wheels. Body in tough polypropylene.
ILM: 15-16 in. (38,1cm-40,6cm)
OL: 34 in. (86,3cm)

TW 3165
CHUBBY

The young 'learner's' delight!—and the perfect first car. Tri-ang toughness and styling comes with the gay red polypropylene body, $6\frac{1}{4}$ in. wheels with duotone hubs, small diameter steering wheel to fit tiny hands—and a wide, extra safe wheelbase. The headlamps are big black eyes!
ILM: 14-15 in. (35,5cm-38,1cm.).
OL: 27 in. (68,5cm.)

Page 6

TM 0106 ROLLS ROYCE

A super luxurious model of the car everyone wants to own. Big-selling points: ultra durable plastic sheet construction on a rugged steel chassis. Dummy hood and windscreen in moulded plastic. Chromed tubular steel bumpers and overriders. Plated rim embellishers. There's a real 6v. printed circuit electrical system—with switch for headlamps and independent switch for side/tail lamps. The $7\frac{1}{2}$ in. wheels have thick long-lasting pvc tyres.
ILM: 16-18 in. (40,6cm.-45,7cm.)
OL: 51 in. (129,5cm.)

TM 0216
PEGASUS

Bang up-to-date in style and colour, with fitted curved windscreen in clear plastic, 6 in. steering wheel with squeaker 'horn', $6\frac{3}{4}$ in. wheels with white laced hubs and chromed hub-caps, adjustable backrest to seat, and sparkling Pegasus decor on the sides and bonnet.
ILM: 15-19 in. (38,1cm.-48,2cm.) OL: 40 in. (101,6cm.)

TM 0725
PANDA PATROL

Kids everywhere think our police cars are wonderful and who can blame them? With chromed radiator, clear curved windscreen, 6 in. steering wheel with squeak-horn, $6\frac{3}{4}$ in. wheels with white moulded plastic hubs. There is a chromed emergency bell and battery operated police-blue light that flashes on and off.
ILM: 16-18 in. (40,6cm.-45,7cm.)
OL: 34 in. (86,3cm.)

TM 0732
BERMUDA

Here's the popular, red, racy roadster to break your sales records. It has a smart white radiator, $6\frac{1}{4}$ in. wheels, plated hub-caps, 6 in. steering wheel with squeak-horn, and a strong curved clear-view windscreen.
ILM: 16-18 in. (40,6cm-45,7cm.).
OL: 34 in. (86,3cm.)

TM 0730
POLICE SQUAD CAR

The young lawman's delight—and handsome too, with its smart navy blue finish, white enamelled radiator, 6 in. steering wheel (with squeak-horn), curved 'clear-view' windscreen and $6\frac{3}{4}$ in. wheels with dashing red moulded hubs. Traffic offenders can be 'gonged' down with the hand-operated police bell.
ILM: 16-18 in. (40,6cm-45,7cm.).
OL: 34 in. (86,3cm.)

Page 2

TM 0140 NEW FOR THE 70's

THUNDERBOLT RACER

An exciting racer to make a flying start into the 70's. Tough blow moulded body with chain drive, dummy engine at rear with bright port inlets and large wheels make this a racer to be proud of. There is a dummy 4 position gear lever and a squeaker button in the steering wheel. ILM: 15-17 in. (38.1 cm — 43,1 cm) OL: 44 in. (111,7 cm)

Page 1

TM 0150 NEW FOR THE 70's

BEACH BUGGIE

Following the trend in versatile all-purpose pleasure runabouts, we are proud to present a super-robust Beach Buggie in pressed steel, with a tough moulded radiator grille and moulded seat. Heavy-duty 9 in. blow-moulded wheels are designed to tackle the toughest terrain. The windscreen folds flat on to the bonnet and the steering wheel is fitted with a horn-squeaker.
ILM: 17 in. - 19 in. (43,1cm - 48,2cm.) OL: 38 in. (96,5cm.)

Page 3

TM 2006
FERGUSON TRACTOR

One of the best selling toy tractors—with its chunky, ultra-rugged body of moulded plastic over a durable steel chassis. Everything's 'realistic', from the engine detail, gear lever and bucket seat, to the 10 in. front and 15 in. rear moulded wheels. Steering wheel has squeak-horn. Adjustable crank drive.
ILM: 19-21 in. (48,2cm.-53,3cm) OL: 37 in. (93,9cm)

NEW FOR THE 70's

TM 2007
PLOUGHMAN TRACTOR

Strong, light and handsome. Tough moulded plastic body with 12 in. blow-moulded rear wheels with strong, extra-grip p.v.c. inserts and 9 in. front wheels. 6 in. steering wheel with squeak-horn. New 'square' front.
ILM: 15-17 in. (38,1cm.-43,1cm.)
OL: 28 in. (71,1cm.)

TM 2300
No. 1 TRAILER

This tough little metal pan, on 5 in. disc wheels, can be towed behind all Tri-ang tractors. With the special sloping end tipping's easy as pie!
OL: 25 in. (63,5cm.)

TM 2009 NEW FOR THE 70's
PRAIRIE TRACTOR

This rugged tractor is in one-piece moulded plastic, with front wheel pedal drive. The 12 in. rear wheels have extra-grip p.v.c. inserts and a moulded front wheel with pvc tyre. 6 in. steering wheel with squeak-horn, and new 'square' front design.
ILM: 14-16 in. (35,5cm.-40,6cm.)
OL: 28 in. (71,1cm.)

TM 2018
TIPPIT

A tough knockabout pedal truck, with a durable pan in moulded plastic that tips up at the touch of a safety-catch lever 7½ in. front and 5⅝ in. rear wheels with colourful disc hubs and 1½ in. chromed hub-caps.
Adjustable cross-bar for:
ILM: 14-16 in. (35,5cm-45,7cm.).
OL: 36 in. (91,4cm.)

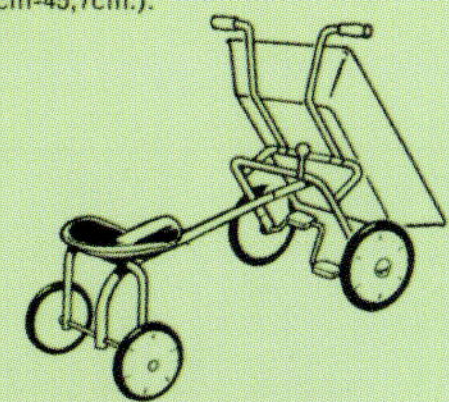

TM 2301
TRAILER No. 2

A useful deep-sided trailer of strong, all-metal construction, with 6½ in. disc wheels. Adds tremendous play value to the Tri-ang tractor range.
OL: 26 in. (66cm.)

TM 2100
DUMPER TRUCK

They'll really go for this tough little earthmover; it has heavy-duty steering, 12 in. moulded plastic wheels with extra-grip inserts, towing eye for trailers and a robust metal frame. The strong metal pan operates with a tipping handle.
ILM: 19-23 in. (48,2cm.-58,4cm.) OL: 44 in. (111,7cm.)

TM 2008
FORD TRACTOR SHOVEL

A first class, highly versatile 'tool' for the budding farmer! It's in robust moulded plastic, with a shaped metal seat and plastic mudguards. The chunky rear wheels have extra-grip inserts twin blow-moulded front wheels, and there's a useful scoop bucket, in moulded plastic on a tubular steel frame, to give unbeatable play value.
ILM: 15-17 in. (38,1cm-43,1cm.)

Tri-ang TM2100 Dumper truck

Tri-ang TW3148 Hovercraft Electric

TM 2016 **FERGUSON TRACTOR (chain drive)**

One of the best selling toy tractors – with its chunky, ultra-rugged body of moulded plastic over a durable steel chassis in bright new colours. Everything's 'realistic', from the engine detail, gear lever and bucket seat, to the 10 in. front and 15 in. rear moulded wheels.
ILM: 19-21 in. (48·2 cm.-53·3 cm.) OL: 37 in. (93·9 cm.)

TM 2007 **PLOUGHMAN TRACTOR**

Strong, light and handsome. Tough moulded plastic body with 12 in. blow-moulded rear wheels with strong, extra-grip p.v.c. inserts and 9 in. front wheels. 6 in. steering wheel with squeak horn. ILM: 15-17 in. (38·1 cm.-43·1 cm.) OL: 28 in. (71·1 cm.)

TM 2010 **PLOUGHMAN ELECTRIC**

Foot down and away you go on this super electric version of the Ploughman Tractor. Powered by rechargeable, battery-operated, 6 volt electric motor.

TM 2300 **No. 1 TRAILER**

This tough little metal pan, on 5 in. disc wheels, can be towed behind all Tri-ang tractors. With the special sloping end tipping's easy as pie!
OL: 25 in. (63·5 cm.)

Page 7

TW 3150 **GT SPORTS**

Everything opens on this eye-opening model, doors, bonnet, and the boot! It's built to last in rugged, brilliant red polypropylene on a strong steel chassis. It has chromed tubular front/rear bumpers, plated radiator and headlamps, and printed dash with a full set of 'instruments'.
ILM: 16-18 in. (40·6 cm.-45·7 cm.) OL: 38 in. (96·5 cm.)

TW 3170 **CARNABY**

Flower-power at its most powerful! Pulsating pink moulded body with contrasting bonnet and boot lid in golden yellow. Duotone disc hubs and fully upholstered seat in "mod" flower p.v.c. completes a car guaranteed to stop the traffic!
ILM: 16-18 in. (40·6 cm.-45·6 cm.) OL: 38 in. (96·5 cm.)

TW 3165 **CHUBBY**

The young learner's delight! – and the perfect first car. Tri-ang toughness and styling comes with the gay red polypropylene body, 6¼ in. wheels with duotone hubs, small diameter steering wheel to fit tiny hands – and a wide, extra safe wheelbase. Also available in bright pink. (Not illustrated). The headlamps are big black eyes!
ILM: 14-15 in. (35·5 cm.-38·1 cm.) OL: 27 in. (68·5 cm.)

Tri-ang TOYS

Page 6

TM 0145 **MILITARY JEEP**

Ideal transport for front-line inspection. Jeep-style body in pressed steel with tough moulded wheels and a fold-flat windscreen and moulded seat. Four-star general's flag is mounted on the moulded radiator.
ILM: 17-19 in. (43·1 cm.-48·2 cm.) OL: 38 in. (96·5 cm.)

TM 0150 **BEACH BUGGIE**

Following the trend in versatile all-purpose pleasure run-abouts is this super-robust Beach Buggie in pressed steel, with a tough moulded radiator grille and moulded seat. Heavy-duty 9 in. blow-moulded wheels are designed to tackle the toughest terrain. The windscreen folds flat on to the bonnet and the steering wheel is fitted with a horn-squeaker.
ILM: 17-19 in. (43·1 cm.-48·2 cm.) OL: 38 in. (96·5 cm.)

Page 4

TW 3115 **KART 'C'**

Tough tubular steel chassis, adjustable padded seat, fast-stopping hand-operated brake on both rear wheels – and a chain drive for go-go-go speed power; and with a low centre of gravity, it's as safe as it's exciting on the corners! Wheels: 7¼ in. front, 10 in. rear, all four with chromed laced hubs and yellow 'knock-off' hub caps.
ILM: 16-18 in. (40·6 cm.-45·7 cm.) OL: 43 in. (109·2 cm.)

TW 3110 **KART 'B'**

'Second cousin' to the Kart 'C', but with adjustable pedal drive and 9 in. rear wheels.

TW 3015 **JUNIOR KART**

Just watch this one go! A safe, robust kart of tubular steel with adjustable bucket seat, and pedal drive, 6½ in. wheels with disc hubs and 1½ in. hubcaps.
ILM: 12-14 in. (30·4 cm.-35·5 cm.) OL: 31 in. (78·7 cm.)

TM 0106 **ROLLS ROYCE**

A super luxurious model of the car everyone wants to own in vacuum-formed plastic sheet on a rugged steel chassis. Chromed tubular steel bumpers and overriders. There's a 6v. printed circuit electrical system – with switch for headlamps and independent switch for side/tail lamps.
ILM: 16-18 in. (40·6 cm.-45·7 cm.) OL: 51 in. (129·5 cm.)

TM 0217 **ASTRA**

Bang up-to-date in style and colour, with new chromed side-rails. The Astra is fitted with curved windscreen, steering wheel with squeaker 'horn', 6¼ in. wheels with white laced hubs and adjustable backrest to seat.
ILM: 15-19 in. (38·1 cm.-48·2 cm.) OL: 40 in. (101·6 cm.)

TM 0725 **PANDA PATROL**

With new chromed side-rails, chromed radiator, clear curved windscreen, 6¼ in. wheels with white moulded plastic hubs. chromed emergency bell and battery operated flashing light.
ILM: 16-18 in. (40·6 cm.-45·7 cm.) OL: 34 in. (86·3 cm.)

TM 0732 **BERMUDA**

Here's the popular racy roadster to break records. It has 6¼ in. wheels, plated hub-caps, and a strong curved clearview windscreen.
ILM: 16-18 in. (40·6 cm.-45·7 cm.) OL: 34 in. (86·3 cm.)

TM 0730 **POLICE SQUAD CAR**

The young lawman's delight, and handsome too, with its smart blue finish, white enamelled radiator, 6 in. steering wheel (with squeak-horn), curved 'clear-view' windscreen and 6¼ in. wheels with dashing white moulded hubs.
ILM: 16-18 in. (40·6 cm.-45·7 cm.) OL: 34 in. (86·3 cm.)

TM 2008 SCOOPAMATIC

In robust moulded plastic, with a shaped metal seat and plastic mudguards. The chunky rear wheels have extra-grip inserts with twin blow-moulded wheels at the front and there's a useful scoop bucket, in moulded plastic on a tubular steel frame, to give unbeatable play value.
ILM: 15-17 in. (38·1 cm.-43·1 cm:)

TM 2009 PRAIRIE TRACTOR

This rugged tractor has a one-piece moulded plastic body in bright dayglo green with front wheel pedal drive. The 12 in. rear wheels and new 7 in. front wheel are all blow moulded and now fitted with extra-grip p.v.c. inserts. The 6 in. steering wheel is fitted with a squeak horn.
ILM: 14-16 in. (35·5 cm.-40·6 cm.) OL: 28 in. (71·1 cm.)

TM 2018 TIPPIT

A tough knockabout pedal truck, with a durable pan in moulded plastic that tips up at the touch of a safety-catch lever. Fitted with large tough moulded wheels with wide extra grip p.v.c. inserts. Adjustable cross-bar for:
ILM: 14-16 in. (35·5 cm.-40·6 cm.) OL: 36 in. (91·4 cm.)

TM 2301 TRAILER No. 2 OL: 26 in. (66 cm.)

TM 2100 DUMPER TRUCK

They'll really go for this tough little earthmover; it has heavy-duty steering, 12 in. moulded plastic wheels with extra-grip inserts, towing eye for trailers and a robust metal frame. The strong metal pan operates with a tipping handle.
ILM: 19-23 in. (48·2 cm.-58·4 cm.) OL: 44 in. (111·7 cm.)

Page 8

TW 3155 SPRITE DE LUXE

Junior Grand Prix drivers will revel in the sleek, low-slung body of moulded polypropylene, with its chromed anti-roll bar, realistic transfers and four rear-mounted exhausts. 7½ in. wheels have plated moulded hubs. Twin rear view mirrors.
ILM: 15-16 in. (38·1 cm.-40·6 cm.) OL: 34 in. (86·3 cm.)

TW 3162 SPRITE

Bound to attract attention on the starting grid is this blue Sprite with its 9 in. blow-moulded wheels. Body in tough polypropylene.
ILM: 15-16 in. (38·1 cm.-40·6 cm.) OL: 34 in. (86·3 cm.)

TW 3145 FORMULA 2 RACING CAR

The most exciting toy racer available, in true Grand Prix style. The tough injection-moulded body in eye-catching red embellished with authentic transfers. The faithful styling includes a 'scoop' air intake and rear-mounted twin exhausts and port inlets. The wheels (7¾ in. front, 8½ in. rear) have strong moulded hubs and are fitted with extra-wide tyres. A real winner!
ILM: 20-22 in. (50·8 cm.-55·8 cm.) OL: 48 in. (121·9 cm.)

Page 5

TM 0140 THUNDERBOLT RACER

An exciting racer to make for flying starts. Tough blow moulded body in bright new dayglo colours. The dummy engine at rear with port inlets and large wheels make this a racer to be proud of. For added play value there is a dummy 4 position gear lever.
ILM: 15-17 in. (38·1 cm.-43·1 cm.) OL: 44 in. (111·7 cm.)

TM 0128 HOVERCRAFT ELECTRIC

The electric battery-powered 'Sit 'n' Ride' toy every boy will want. Smooth strong plastic – moulded in a modern dayglo colour, with stop-go foot-operated button. Single front-wheel steering provides fast about-turns in dodg'em car fashion. The battery is non-spill and re-chargeable. Seat height 8½ in.
OL: 23 in. (58·4 cm.)

Tri-ang TOYS

TW 3172 FUN BUGGIE

Every youngster with an eye for something that's different will clamour for this new, exciting Tri-ang Fun Buggie. In tough injection-moulded polypropylene, mounted on a rugged steel frame with heavy duty wheels at each corner. Headlights add a luxury touch.
ILM: 16-17 in. (40·6 cm.-43·2 cm.) OL: 34 in. (86·4 cm.)

Tri-ang TOYS

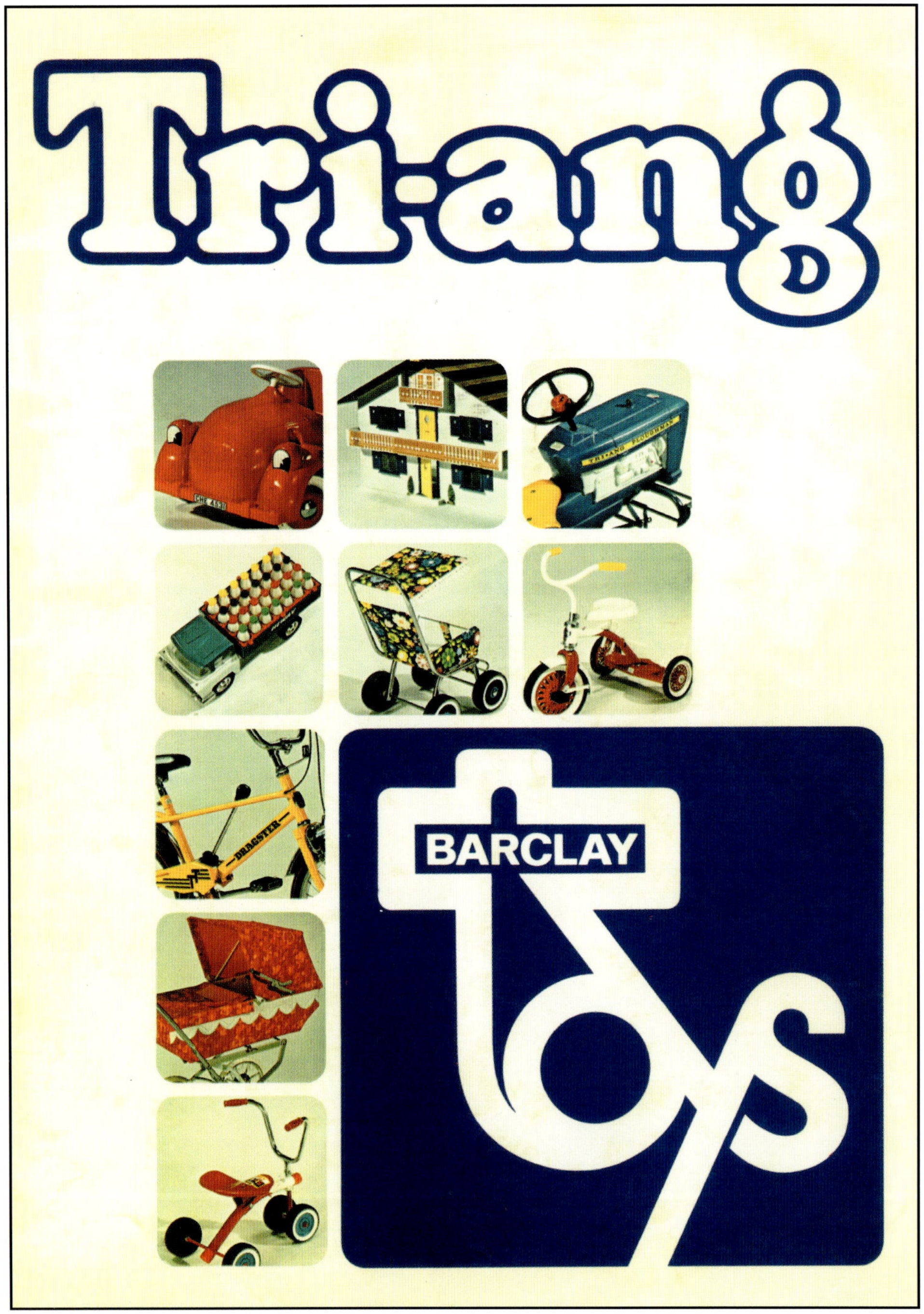

TW 3155 SPRITE DE LUXE

Junior Grand Prix drivers will revel in the sleek, low-slung body of moulded polypropylene, with its chromed anti-roll bar, realistic transfers and four rear-mounted exhausts. 7½ in. wheels have plated moulded hubs. Twin rear view mirrors. ILM: 15-16 in. (38·1 cm.-40·6 cm.) OL: 34 in. (86·3 cm.)

TW 3162 SPRITE

Bound to attract attention on the starting grid is this blue Sprite with its 9 in. blow-moulded wheels. Body in tough polypropylene. ILM: 15-16 in. (38·1 cm.-40·6 cm.) OL: 34 in. (86·3 cm.)

TW 3165 CHUBBY

The young learner's delight! – and the perfect first car. Tri-ang toughness and styling comes with the gay red polypropylene body, 6¼ in. wheels with duotone hubs, small diameter steering wheel to fit tiny hands – and a wide, extra safe wheelbase. The headlamps are big black eyes! ILM: 14-15 in. (35·5 cm.-38·1 cm.) OL: 27 in. (68·5 cm.)

Page 4

TW 3152 **RALLY CAR**

As rugged as they come – and built to last. That's the new Tri-ang Rally Car with its polypropylene body, mounted on a strong steel chassis and fitted with 7½ in. heavy-duty blow-moulded wheels with hard-wearing p.v.c. tyres. As well as having plated 'headlamps' on the body it has a set of four 'long-range driving lamps' mounted on a wire grille in true rally car style. There's an anti-roll bar fitted and a spare wheel mounted on the opening boot. The bonnet opens to reveal a dummy engine – what play value there is in this exciting model!
ILM: 16-18 in. (40,6cm.-45,7cm.) OL: 38 in. (96,5cm.)

TW 3144 **FORMULA 3 RACING CAR**

In true Grand Prix style, the Formula 3 is updated with the addition of an aerofoil over the rear wheels. The tough injection-moulded body in eye-catching red is embellished with authentic transfers. The faithful styling includes a 'scoop' air intake and rear-mounted twin exhausts. Tough blow-moulded wheels (7½ in. front, 9 in. rear) have hard-wearing p.v.c. tyres. A real winner!
ILM: 20-22 in. (50,8cm.-55,8cm.) OL: 48 in. (121,9cm.)

Tri-ang TOYS

MILITARY JEEP

BEACH BUGGIE

TM 0145 **MILITARY JEEP**

Ideal transport for front-line inspection. Jeep-style body in pressed steel with tough moulded wheels and a fold-flat windscreen and moulded seat. Four-star general's flag is mounted on the moulded radiator.
ILM: 17-19 in. (43·1 cm.-48·2 cm.) OL: 38 in. (96·5 cm.)

TM 0150 **BEACH BUGGIE**

Following the trend in versatile all-purpose pleasure run-abouts is this super-robust Beach Buggie in pressed steel, with a tough moulded radiator grille and moulded seat. Heavy-duty 9 in. blow-moulded wheels are designed to tackle the toughest terrain. The windscreen folds flat on to the bonnet and the steering wheel is fitted with a horn-squeaker.
ILM: 17-19 in. (43·1 cm.-48·2 cm.) OL: 38 in. (96·5 cm.)

PANDA PATROL

BERMUDA

FUN BUGGIE

TM 0725 **PANDA PATROL**

The young lawman's delight, and handsome too, with its smart blue finish, chromed radiator, windscreen, 6¼ in. wheels with white moulded plastic hubs. There is a chromed bell for emergencies, and a battery-operated flashing light. ILM: 16-18 in. (40,6cm.-45,7cm.) OL: 34 in. (86,3cm.)

TM 0732 **BERMUDA**

Here's the popular racy roadster to break records. It has 6¼in. wheels, plated hub-caps, and a strong curved clearview windscreen.
ILM: 16-18 in. (40,6cm.-45,7cm.) OL: 34 in. (86,3cm.)

TW 3172 **FUN BUGGIE**

Every youngster with an eye for something that's different will clamour for this new, exciting Tri-ang Fun Buggie. In tough injection-moulded polypropylene, mounted on a rugged steel frame with heavy duty wheels at each corner. Headlights add a luxury touch. ILM: 16-17 in. (40,6cm.-43,2cm.) OL: 34 in. (86,4cm.).

No. 1 TRAILER

PLOUGHMAN TRACTOR

SCOOPAMATIC

PRAIRIE TRACTOR

TM 2007 **PLOUGHMAN TRACTOR**

Strong, light and handsome. Tough moulded plastic body with 12 in. blow-moulded rear wheels with strong, extra-grip p.v.c. inserts and 9 in. front wheels. 6 in. steering wheel with squeak horn. ILM: 15-17 in. (38·1 cm.-43·1 cm.) OL: 28 in. (71·1 cm.)

TM 2300 **No. 1 TRAILER**

This tough little metal pan, on 5 in. disc wheels, can be towed behind all Tri-ang tractors. With the special sloping end tipping's easy as pie!
OL: 25 in. (63·5 cm.)

TM 2008 **SCOOPAMATIC**

In robust moulded plastic, with a shaped metal seat and plastic mudguards. The chunky rear wheels have extra-grip inserts with twin blow-moulded wheels at the front and there's a useful scoop bucket, in moulded plastic on a tubular steel frame, to give unbeatable play value. ILM: 15-17 in. (38,1 cm.-43,1 cm.).

TM 2009 **PRAIRIE TRACTOR**

This rugged tractor has a one-piece moulded plastic body with front-wheel pedal-drive. The 12 in. rear wheels and 7 in. front wheel are blow-moulded and fitted with extra-grip p.v.c. tyres. The 6 in. steering wheel is fitted with a squeaker horn. ILM: 14-16 in. (35,5cm.-40,6cm.) OL: 28 in. (71,1cm.)

TW 3120 **KART "D"**

A new Tri-ang Kart designed for robustness and play value. It has a really tough tubular steel frame with 9 in. strong blow-moulded wheels. There is a hand-operated two-wheel brake and an adjustable padded seat. Chain-driven for top speeds.
ILM: 18-22 in. (45,7cm.-55,9cm.) OL: 43½ in. (110,5cm.)

TW 3107 **KART "A"**

Just watch this one go! A safe, robust kart of tubular steel with adjustable bucket seat and pedal drive. It has 7½ in. blow-moulded wheels with p.v.c. tyres.
ILM: 12-14 in. (30,4cm.-35,5cm.) OL: 33 in. (83,8cm.)

TW 3110 **KART "B"**

This is an established favourite on the karting circuits. It has a tough steel frame and adjustable pedal drive for instant get-up-and-go. The wheels have chromed laced hubs with "knock-off" hub-caps.
ILM: 16-18 in. (40,6cm.-45,7cm.) OL: 43 in. (109,2cm.)

Page 7

TW 3720 **TIPPIT**

A tough knockabout pedal truck, with a durable pan in moulded plastic that tips up at the touch of a safety-catch lever. Fitted with large tough moulded wheels with wide extra grip p.v.c. inserts. Adjustable cross-bar for: ILM: 14-16 in. (35·5 cm.-40·6cm.) OL: 36 in. (91·4 cm.)

TM 2301 **TRAILER No. 2** OL: 26 in. (66 cm.)

TM 3710 **DUMPER TRUCK**

They'll really go for this tough little earthmover; it has heavy-duty steering, 12 in. moulded plastic wheels with extra-grip p.v.c. tyres, towing eye for trailers and a robust metal frame. The strong metal pan operates with a tipping handle. ILM: 19-23 in. (48·2 cm.-58·4 cm.) OL: 44 in. (111·7 cm.)

TM 2020 **JUNIOR FARMER**

Great fun for a young farm hand. Strong blow-moulded tractor with 7½ in. blow-moulded wheels with p.v.c. tyres. ILM: 12-15 in. (30·5 cm.-38·1 cm.) OL: 22½ in. (57·2 cm.)

Page 6

Mobo

BARCLAY toys

BARCLAY toys

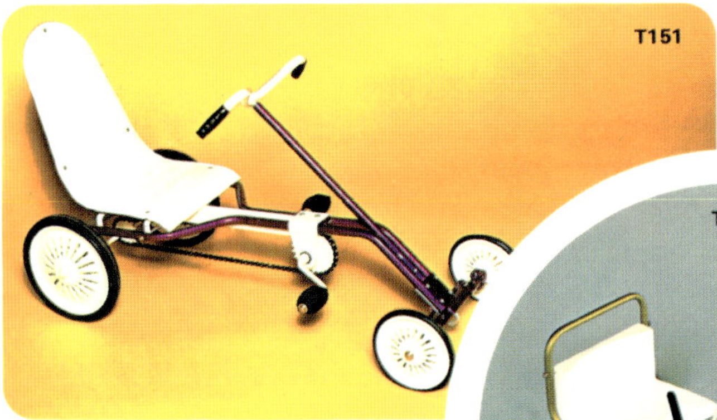

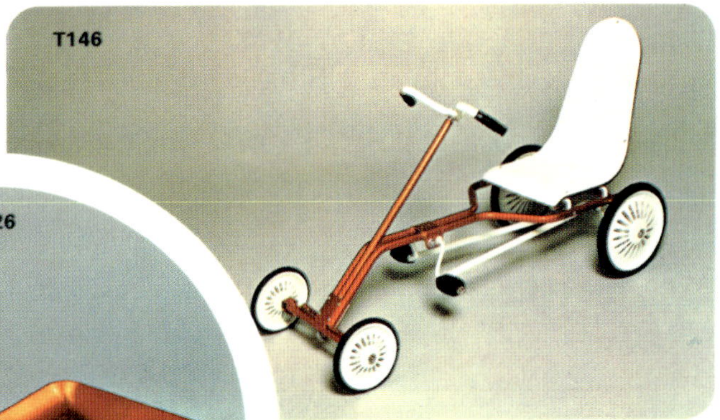

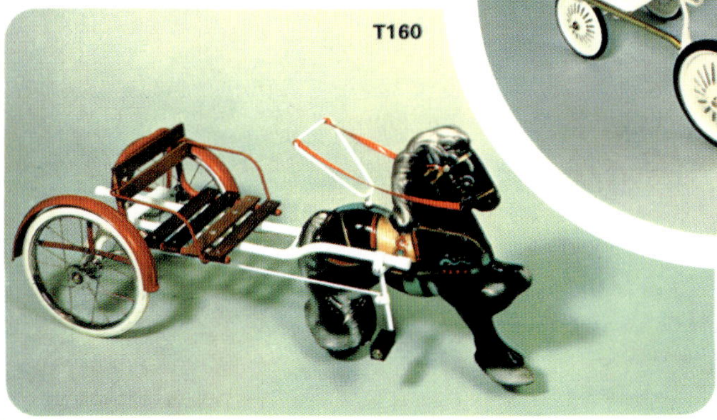

T130 4-WHEEL TRACTOR
A smart modern tractor that will delight all budding young farmers. An extremely strong, blow-moulded body with realistic detail, blow-moulded plastic wheels with rubber tyres. Plastic steering wheel and crank pedal drive.
Leg reach 14" (35.6 cms). Length $28\frac{1}{2}$" (72.4 cms).
Width $19\frac{3}{4}$" (50.2 cms). Height $22\frac{3}{4}$" (57.8 cms).

T135 3-WHEEL TRACTOR
With the same strong blow-moulded body as the 4-wheel version. Front wheeled drive – approx. 17" (43.3 cms) leg reach. 8" (20.3 cms) front, $11\frac{1}{4}$" (28.6 cms) rear, blow-moulded wheels with rubber tyres. Length $28\frac{1}{2}$" (72.4 cms).
Width $16\frac{1}{2}$" (41.9 cms). Height $22\frac{1}{2}$" (57.2 cms).

T151 DEMON RACER GO KART
A nippy chain drive Go Kart, with padded one-piece seat and backrest. Colourful in flamboyant purple and white stoved enamel. 'Festival' pattern wheels, 10" (25.4 cms) rear, $6\frac{1}{2}$" (16.5 cms) front. Nylon bearings.
Leg reach. 15"-17" (38.1-43.2 cms) Length 40" (101.6 cms).
Width 18" (45.7 cms). Height of backrest 22" (55.9 cms).

T146 SCORCHER GO KART
A less expensive version of the Demon Racer, with crank-drive, gold and white stoved-enamel finish.
'Festival' pattern wheels, 10" (25.4 cms) rear, $6\frac{1}{2}$" (16.5 cms) front. Plastic handgrips. Leg reach 15"-17" (38.1-43.2 cms).
Length 40" (101.6 cms). Width 18" (45.7 cms).
Height of backrest 22" (55.9 cms).

T145 TRACKMASTER
Practically every moving part has nylon bearings for long trouble-free running. Specifications: Sturdy $\frac{7}{8}$" (2.3 cms) steel tube frame. O/D Front tyres 8" dia. x $1\frac{1}{4}$" wide. (20.3 x 3.3 cms). Rear tyres 9" dia. x 2" wide (22.9 x 5.1 cms). Nylon bushed king-pins and correct toe-in on the front wheels give top accuracy steering. Seat is body moulded and angled for more thrust, more powerful pedal action, covered in washable foam-padded P.V.C., it adjusts for children aged 5 years and upwards. Overall length $42\frac{1}{2}$" (108 cms). Width 19" (48.3 cms). Height to top of steering wheel $21\frac{3}{4}$" (55.3 cms).

T126 TIPPER TRUCK
For all dirt loving youngsters! A colourful tipper truck with easy joystick steering, and front wheel pedal drive. Pull back the right-hand lever, and the pan will dump its load. Plastic moulded seat and backrest. Front wheels $9\frac{1}{2}$" dia. (24.1 cms). Rear wheels $6\frac{1}{4}$" dia. (15.9 cms). Leg reach 15" (38.1 cms). Length 32" (81.3 cms). Width 17" (43.2 cms). Height $15\frac{1}{2}$" (39.4 cms).

T160 PONY CART
A splendid pedal toy. With slatted wooden seat and rubber-tyred wheels. Easily steered left or right. Wheels 12" x $1\frac{3}{8}$" (30.5 x 3.5 cms). Length 42" (106.7 cms). Adjustable leg reach 15"-19" (38.1-48.3 cms).

BARCLAY toys

T920 MATRA FORMULA I CAR
A car which every boy dreams of owning. A racing car to beat all others! Low slung body formed from strong royal blue plastic – deep snug driver's seat – plastic steering wheel with central hooter, and eye-catching racing transfers.
Leg reach approx. 15½"-17" (39.4-43.2 cms).
Length 47" (119.4 cms). Width 27" (68.6 cms).
Height 14" (35.5 cms).

T917 PORSCHE CAR
A zippy looking sports car, which will give youngsters plenty of thrills. Moulded from strong red plastic, it has a modern one-piece steering wheel into which a hooter is set. Detachable hood. Realistic body detail. 7½" (19 cms) dia. plated spoked wheels. Leg reach 16½"-18" (41.9-45.7 cms). Length 44" (111.8 cms). Width 19" (48.3 cms).

T914 ALFA ROMEO CAR
A smart little tourer, inexpensively priced, formed from strong plastic. The front wheel drive allows the younger rider to trundle, as well as pedal it. Plated wheels, one-piece steering wheel. Leg reach 15" (38.1 cms). Length 32" (81.3 cms). Width 15" (38.1 cms).

T915 MINI-FORD CAR
Similar to the Alfa Romeo, but with slightly different body moulding. Leg reach 15" (38.1 cms). Length 32" (81.3 cms). Width 15" (38.1 cms).

T103 POLICE PATROL CAR
Attractively priced, it has a flashing warning light, disc wheels and a plastic windscreen. Long wearing nylon bearings throughout. Leg reach 16"-19" (40.6-48.3 cms).
Length 34¾" (88.3 cms). Width 14½" (36.8 cms).
Height 16½" (41.9 cms).

T102 POPULAR PEDAL CAR
A popular car at a popular price. Bright red stoved enamel finish. Plastic windscreen and steering wheel. Rubber tyred disc wheels. Leg reach adjustable 16"-19" (40.6-48.3 cms).
Length 34¾" (88.3 cms). Width 14½" (36.8 cms).
Height 16½" (41.9 cms).

T904 RACING HELMET AND GOGGLES
Tough plastic helmet – padded and lined, adjustable chin strap. Goggles hinged across the nose-piece, elastic headstrap.
Available in U.K. and Northern Ireland only.

T917

T914

T915

T920

T103

T102

T904

PEDAL CARS

106001 CHUBBY

The perfect first car! Safe, full of character and tough enough to look after itself. Bright red polypropylene body, with bright $6\frac{1}{4}$ in wheels, and a small diameter steering wheel for tiny hands. With a new, appealing 'Noddy' car look for 1973. Inside leg measurement: 14-15 ins (35.5 cms). Overall length: 27 ins (68.5 cms).

Britain's best selling pedal car. A real favourite with 2-4 year olds now with a new low price to make it even more popular with parents.

106004 FUN BUGGY

Super, low slung 'buggie' lines in tough injection moulded polypropylene mounted on a rugged steel frame. Heavy.duty wheels with a wide, safe wheel span. Headlights add a luxury touch. Inside leg measurement: 16-17 ins (40.6-42.3 cms). Overall length: 34 ins (86.4 cms)

Best selling pedal car in Britain for the 4-7 year olds. And the new 1973 price brings it into the biggest volume price bracket for pedal cars.

PEDAL CARS

106002 SPRITE

A real 'vroom' car for tomorrow's Grand Prix drivers. This blue, formula racer is made in tough polypropylene with big 9 inch, wide slung racing wheels. Inside leg measurement: 15-16 ins (38.1 cms). Overall length: 34 ins (86.3 cms).

A traditional top seller. The only injection moulded pedal racing car in the best selling £6-£8 price group.

106007 BERMUDA

Popular, low priced sports car model in bright red with white trim. $6\frac{1}{4}$ inch wheels, plated hub caps and strong clearview windscreen. Inside leg measurement: 16-18 ins (40.6-45.7 cms). Overall length: 34 ins (86.3 cms).

The only reasonably priced metal pedal car on the market. This model will give years of play value

106008 PANDA PATROL

All the realism of a motorway patrol car. Complete with chromed emergency bell, battery operated flashing light, clearview windscreen, chromed radiator and $6\frac{1}{4}$ inch wheels. Inside leg measurement: 16-18 ins (40.6-45.7 cms). Overall length 34 ins (86.3 cms).

Every child wants the authority, excitement and action of the police driver's life, which is why this is such a top selling model.

PEDAL CARS

106003 CHASER

A bright new, streamlined fun car for 3-6 year olds with the added appeal of sweeping 'dragster' lines. The racing wheels with full width tyres are bigger at the back than at the front – giving it a low slung front appearance. Made from tough injection moulded polypropylene with 'look-out' eyes for headlights. Inside leg measurement: 15 ins (38.1 cms). Overall length: 33 ins (83.7 cms).

A certain best seller for 1973. This super new addition combines the advantages of spanning both the best selling price range and the best selling age bracket.

106006 CHARGER

A real souped-up addition to the pedal car range! Exciting, sporty roadster lines in brilliant white. With new pedal action, wide racing wheels, full width tyres, independent spring suspension and ASA super plastic moulded body. Inside leg measurement: 13 ins (33.1 cms). Overall length: 42 ins (105.2 cms).

This new model is aimed at the big 20% of the market for the £10 and over price range. It also has the advantage of being a sit-on rather than a sit-in model – thus giving it a greater age range.

Tri-ang
The nice part of growing-up
NEW
NEW
NEW
NEW
Catalogue
1973

GO-KARTS

103002 INDIANAPOLIS 500
Top speed chain driven kart for the older age group. A really robust model with tough tubular steel frame, finished in new flamboyant red and silver. New wide racing wheels with full width tyres and two wheeled hand brake give extra road holding and stopping power. And the new padded seat gives extra comfort. Now with new chain tensioner. Inside leg measurement : 18-22 ins (45.7-55.9 cms). Overall length : 43½ ins (110.5 cms).
A much improved version of the biggest selling chain driven go-kart in Britain.

103002

103001 GROUNDKART
A wide, squat, brightly coloured, pedal driven kart for the 4-7 year olds. This now has a really tough construction with great chunky wheels, full width tyres and a generous, blow moulded steering wheel. Inside leg measurement : 12-16 ins (30.4-35.5 cms). Overall length : 33 ins (83.3 cms).
The improvements to last year's popular Kart A make this the top value kart for the 4-7 year olds and a certain best seller this year in its class.

103004

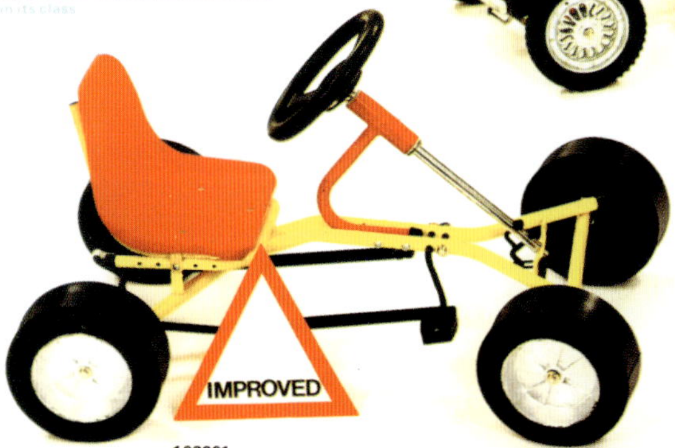

103001

103004 MONACO
A super looking new model for the 5-11 age group in eye catching red, yellow and black with racing numbers. Spoked wheels with fat, deep tread tyres, high speed chain drive, two wheel hand brake, plastic moulded seat, front cowling, chain guard and high rise, 'easy rider' back bar. Inside leg measurement : 16 ins (40.6 cms). Overall length : 43 ins (109.2 cms).
All set to become Europe's biggest selling kart in an expanding go-kart market.

103003

103003 MINI KART
A lot of kids have been waiting for this one ! A mini-kart for children in the 2-4 age group. A really cheeky little kart with all the toughness and styling of the bigger models. Bright red safety bucket seat, fat blow moulded wheels with full width tyes. Inside leg measurement : 13 ins (33.1 cms). Overall length : 29 ins (73.5 cms).
This is an important new addition expanding the go-kart market to include the younger 2-4 year olds.

PEDAL CARS

106009 PINK PANTHER
A sensational new model based on the well known Pink Panther car from the T.V. series. Exciting, futuristic design, with foot operated steering and unique arm drive lever action. The only rotationally moulded car on the market – with extra strength throughout every part of its colourful body. A car with the looks and innovations to capture any child's imagination this year.
Pre-sold character merchandising comes to wheeled toys. The Pink Panther car will revolutionise and expand the whole pedal car market.

106003 CHASER
A combination of amusing personality and dragster 'zap'. Chaser must be *the* 'in' car of the 70's. Inside leg measurement: 15 ins (38.1 cms). Overall length: 33 ins (83.1 cms).

106004 FUN BUGGY
Low and wide, with its gleaming headlamps, the Fun Buggy is set for even better sales this year. Inside leg measurement: 16 – 17 ins (40.6 – 43.2 cms). Overall length: 34 ins (86.4 cms).

106007 BERMUDA
This snappy sports car is one of the most competitively priced metal cars on the market. Inside leg measurement: 16 – 18 ins (40.6 – 45.7 cms). Overall length: 34 ins (86.3 cms).

106001 CHUBBY
The perfect first car! It's tough, it's safe, it's full of character. And even the steering wheel is just right for a tiny driver. Inside leg measurement: 14 – 15 ins (38.1 cms). Overall length: 27 ins (68.5 cms).

106002 SPRITE
Long, low and stylish, the Sprite captures the imagination of any budding racing driver. And its tough polypropylene construction can withstand plenty of punishment. Inside leg measurement: 15 – 16 ins (40.1 cms). Overall length: 34 ins (86.3 cms).

Tri-ang
Go-Karts

103001 GROUNDKART
Tough and chunky with low, wide looks. A big favourite with the 4 – 7 age group. Inside leg measurement: 12 – 16 ins (30.4 – 40.1 cms). Overall length: 33 ins (83.3 cms).

103002 INDIANAPOLIS 500
Here's one that really goes – especially in the showroom. Dozens of exciting features make this high-performance kart a must for older children. Inside leg measurement: 18 – 22 ins (45.7 – 55.9 cms). Overall length: 43½ ins (110.5 cms).

103003 MINI KART
At last, a really exciting kart for the 2 – 4 age group, featuring all the long-lasting appeal of the bigger Tri-ang karts. Inside leg measurement: 13 ins (33.1 cms). Overall length: 29 ins (73.5 cms).

Tri-ang
Sit 'n' Ride

101003 STRIDA RACER
With its extra wide racing wheels and squat lines, this big seller looks racier than ever. While its low centre of gravity means that it's very safe. Seat height: 8¼ ins (20.9 cms). Overall length: 30 ins (76.2 cms).

101002 GROUNDHOG
A superb character toy that brings the ever-popular Groundhog shape into 3-dimensional life. Seat height: 8½ ins (21.6 cms). Overall length: 20 ins (50.8 cms).

Tri-ang
Tractors

107001 PLOUGHMAN TRACTOR
Tough, realistic and competitively priced, the Ploughman features a very comfortable seat.
Inside leg measurement: 15 – 17 ins (38.1 – 43.1 cms). Overall length: 28 ins (71.1 cms).

107006 TRI-TRAILER
Tough enough to carry all sorts of heavy loads, this colourful trailer hitches onto all Tri-ang Tractors. Overall length: 26 ins (66 cms).

107011 FIELDMASTER
A smart realistic-style tractor, with pedal drive and all the strength necessary for heavy work on the land. Inside leg measurement: 14 – 16 ins (35.5 – 40.6 cms).
Overall length: 28 ins (71.1 cms).

107004 SCOOPAMATIC
Featuring a sturdy loader bucket, this is one of Britain's most exciting tractors. It's a sure seller! Inside leg measurement: 15 ins (38.1 cms).

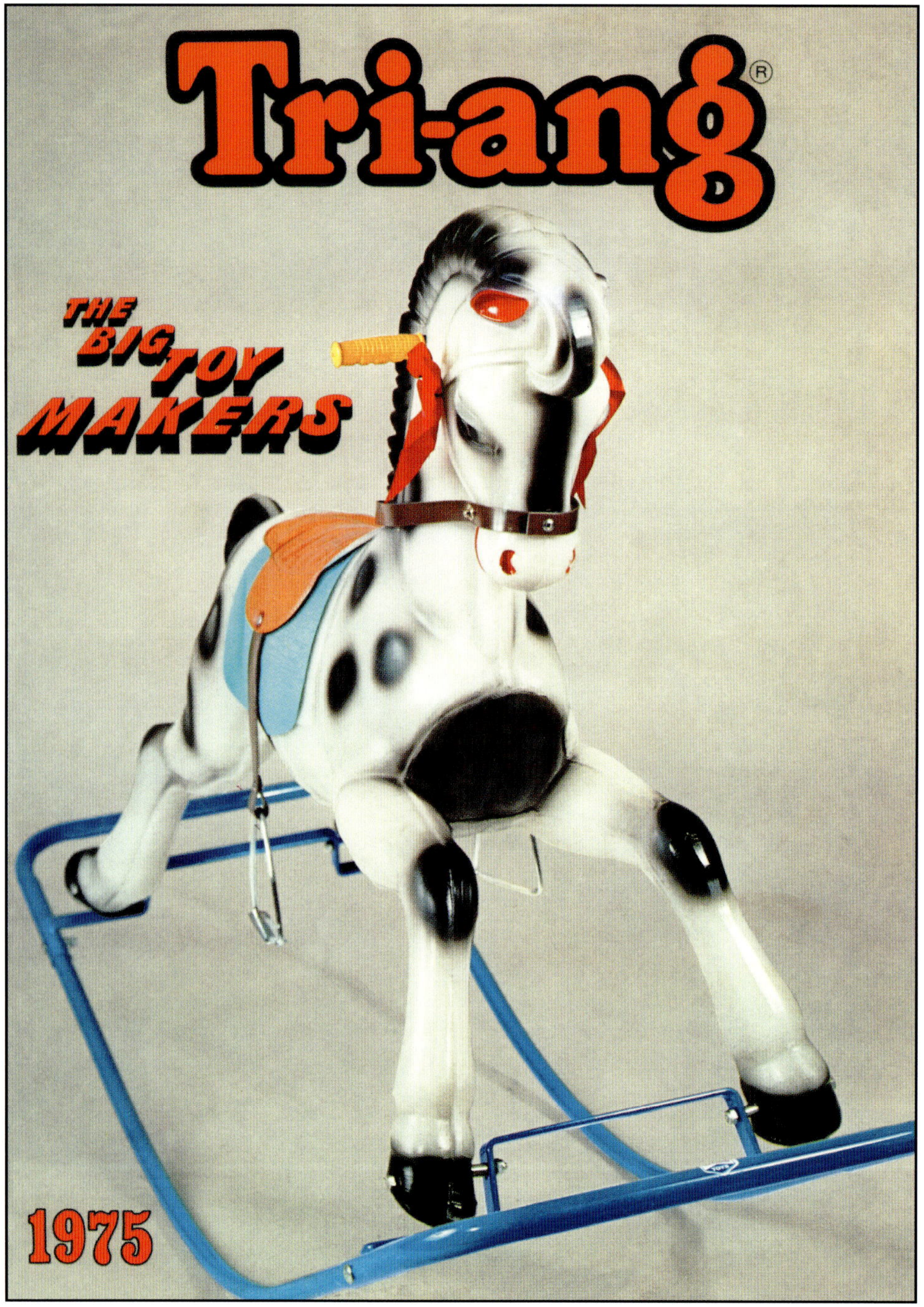

103003
Minikart

For the 2–4 age group. Inside leg measurement 13 ins. (33·1 cms). Total length 29 ins. (73·5 cms).

103001
Groundkart

For the 4–7 age group. Inside leg measurement 12–16 ins. (30·4–40·1 cms). Total length 33 ins. (83·3 cms).

103004
Silverstone Go-Kart

new

Chain Driven. Inside leg measurement 16–20 ins. (40·6–53·5 cms). Total length 37 ins. (94 cms) 25 ins. width (63·5 cms).

103002
Indianapolis

For older children. Inside leg measurement 18–22 ins. (45·7–55·9 cms). Total length 43½ ins. (110·5 cms)

Tri-ang Toys
From M.V. Tri-ang
MOON/FUN BUGGY
Ref. 4005
Great value as well as great fun, this is two cars in one. Two sets of transfers come with the car so it can be a Moon Buggy or a Fun Buggy. For children aged 4 and above.
JEEP RENEGADE
Ref. 4015
A great action car for 3 year olds and above, the Jeep has an opening bonnet and roll over bar.
Then, if it's more than pedal power that you want, a controlled speed motorised version is available powered by a rechargeable 6v battery. It too has an opening bonnet and roll over bar. Suitable for the 3 years plus age group.
CHASER POLICE CAR
Ref. 4009
One of the most exciting Tri-ang pedal cars, equipped with a flashing blue light and siren. Suitable for children aged 3 and above.
FIRE CHIEF
Ref. 4007
A realistic bright red car with a working flashing blue light. Suitable for the 4 years plus age group.
POLICE
FIRE CHIEF
CHUBBY CAR
Ref. 4000
For the child who can't wait to get behind the wheel, the Chubby is the first in the Triang range of pedal cars, specially for the 2–4 year age group.
SPRITE
Ref. 4003
For the James Hunt of the year 2000, a Grand Prix racing car for the 4–6 year olds.

Tri-ang Toys
From M.V. Tri-ang
BABYWALKER
Ref. 4700–05
A traditional stable Babywalker, available on its own or with colourful plastic or wooden bricks. Ideal for a toddler's first steps.
TRACTOR
Ref. 4400
The perfect toy for the prospective farmer, the tractor features chain drive, realistic steering, moving switch controls and a body moulded in polypropylene.
And for extra versatility, this trailor attachment can be fitted. Ref. 4401.
Made as strong as the tractor, it will carry really heavy loads.
Suitable for children age 4 and above.
FORD
7700
WHEELBARROW
Ref. 4405
A play wheelbarrow for very small children.

Tri-ang Price Guide

This price guide uses actual Tri-ang catalog reference numbers (where appropriate) and descriptions. With regard to the price valuations given, these are based upon market values and auction prices, and the values placed upon collectors', dealers', and traders' cars in present or past situations. The prices quoted are as accurate and up-to-date as possible based on the sources that provided them, bearing in mind the condition, rarity, and other prevailing factors of demand and supply, along with the market situation at the time of writing. However, the author and the contributors to this book take no responsibility for any action taken in buying, selling, or trading of cars by anyone consulting this book; it is intended only to be a guide.

1953 -1954

6001B-£450-£500; $650-$750
6002A-£350 £400; $500-$600
6003 -£400-£450; $600-$650
6004A-£250 £300; $400-$450
6005B-£250-£300; $400-$450
6006A-£150-£200; $200-$300
6007B-£150-£200; $200-$300
6008 -£250-£300; $400-$450
6009 -£400-£450; $600-$650
6010 -£400-£450; $600-$650
6011 -£400-£450; $600-$650
6012 -£350-£400; $500-$600
6013 -£300-£350; $450-$500
6014 -£150-£200; $200-$300
6015 -£100-£150; $150-$200
6016 -£250-£300; $400-$450
6017 -£350-£400; $500-$600
6018 -
6019 -£350-£400; $500-$600
6020 -£900-£1,000; $1,300-$1,450
6021 -£100-£150; $150-$200
6022 -£100-£150; $150-$200
6023 -£150~£200: $200-$300
6024 -

1955-1956

7001 -£850-£950; $1,250-$1,400
8001 -£1,250-£1,500; $1,800-$2,200
8002 -£250-£350; $400-$500
7000 -£450-£500; $650-$750
7601 -£350-£450; $500-$650
6002 -£350-£450; $500-$650
7003 -£450-£500; $650-$750
8003 -£350-£450; $500-$650
8004 -£450-£500; $650-$750
6003 -£400-£450; $600-$650
6013 -£350-£400; $500-$600
7007 -£200-£250; $300-$400
7606 -£200-£250; $300-$400
7009 -£150-£200; $200-$300
6007 -£150-£200; $200-$300
6006 -£150-£200; $200-$300
8005 -£300-£350; $450-$500
6023 -£250-£300; $400-$450
6016 -£300-£350; $450-$500
6017 -£350-£400; $500-$600
7020 -£100-£150; $150-$200
8006 -£200-£250; $300-$400
6021 -£100-£150; $150-$200
6009 -£450-£500; $650-$750
8007 -£650-£700; $950-$1,000
6011A-£400-£450; $600-$650
8008 -£150-£200; $200-$300
6014 -£200-£250; $300-$400
7611 -£300-£350; $450-$500

1961

Morris Oxford to Austin A60 -£350-£450; $500-$650
T.60 Police car -£350-£400; $500-$600
T.65 -£250-£300; $400-$450
T. 50 -£200-£250; $300-$400
T. 45C Miami -£450-£500; $650-$750
Triumph Herald "B" -£500-£550; $750-$850
Triumph Herald "A" -£450 £500; $650-$750
T.45.A -£150-£200; $200-$300
T.45.B -£200-£250; $300-$400
T 40B Monte Carlo -£200-£250; $300-$400
T.30 Tipper -£250-£300; $400-$450
T. 30 Ambulance -£300-£350; $450-$500
T. 30 Fire Chief -£350-£400; $500-$600
T. 30 Police Car -£300-£350; $450-$500
T.30A Thunderbolt -£150-£200; $200-$300
T. 30B Lightning -£200-£250; $300-$400
Jeeps -£250-£350: $400-$500
Noddy Car -£200-£250; $300-$400
Royal Prince -£150-£200; $200-$300
Duke -£100-£150; $150-$200
Comet "B" -£150-£200; $200-$300
Comet "A" -£100-£150; $150-$200
Formula 1 -£400-£450; $600-$650
Grand Prix -£300-£350; $450-$500
M.G.S. -£450-£500; $650-$750
M.G.A. -£300-£350; $450-$500
Super tractor -£350-£400; $450-$600
Midget tractor -£150 £200; $200-$300
Super go-kart -£200-£250; $300-$400

Junior go-kart "B" -£150-£200; $200-$300

Supersonic plane -£450-£500; $650-$750

1962

Veteran motor car -£250-£300; $400-$450

Wooden pedal engine "prairie chief" -£650-£700; $950-$1,000

Others repeated and as priced in previous catalogue

1963

Jaguar Mk X -£350-£400; $500-$600

M.G.Midget -£250-£350; $400-$500

Ferguson tractor -£150-£200; $200-$300

Harvester tractor -£150-£200; $200-$300

Farmer's boy tractor -£100-£150; $150-$200

Midget tractor -£100-£150; $150-$200

1964

Rolls-Royce -£450-£500; $650-$750

"E" Type Jaguar -£300-£350; $450-$500

Triumph TR4 -£350-£400; $500-$600

Others repeated and as priced in previous catalogue

1965

Junior Grand Prix Racer -£200-£250; $300-$400

"E"-type Jaguar with motor roar unit -£650-£700; $950-$1,000

Junior Grand Prix Racer with motor roar unit -£500-£550; $750-$850

1967

"Blueblrd" pedal boat -£750-£800; $1,100-$1,200

Safari Overlander with trailer -£550 £600; $850-$900

Corsair -£350-£400; $500-$250

Traveller -£250-£300; $400-$450

Police radio car -£250-£300; $400-$450

Clubman -£200-£250; $300-$400

Gadabout -£150-£200; $200-$300

Fire service truck -£350-£400; $500-$600

TM 0705-TM 0731 average -£250-£350; $400-$500

Others repeated and as priced in previous catalogue

1968

"Sea Spray" pedal boat -£850 £900; $1,250-$1,300

TW 0900- TWO908-average -£100~£200; $150-$300

TM 0136 Moon Rocket -£750-£850; $1,100-$1,250

TM 1520 Armoured car -£250-£350; $400-$500

TM 1535 Jungle camp set -£350-£400; $500-$600

Others repeated and as priced in previous catalogue

1970

TM 0140 Thunderbolt racer -£100-£150; $150-$200

TM 0150 Beach buggie -£100-£150; $150-$200

TW 3146 Formula 4 (electric model) -£350-£400; $500-$600

TW 3145 Formula 2 racing car -£300-£350; $450-$500

TW 3148 Hovercraft electric -£100-£150; $150-$200

TW 3170 Carnaby -£200-£250; $300-$400

TW 3150 GT sports -£350-£400; $500-$600

TW 3155 Sprite de luxe -£250-£300; $400-$450

TW 3162 Sprite -£150-£200; $200-$300

TW 3165 Chubby -£50-£75; $70-$100

TM 2006-2300-average -£100-£150; $150-$200

TM 2018-Tippit -£50-£75; $70-$100

TM 2100 Dumper truck -£100-£150; $150-$200

TM 2301 -Trailer No.2 -£50~£75; $70-$100

TM2008-Tractor shovel -£150-£200; $200-$300

Go-karts and others repeated and as priced in previous catalogue

1971

TW 3172 Fun buggie -£75-£100; $100-$150

No other new items for this catalogue, others repeated and priced in previous catalogue.

1972

Mobo Barclay Toys (D. Sebel Ltd.) taken over by Tri-ang in this year.

T920-Matra Formula 1 car -£200-£250; $300-$400

T917-Porsche car -£100-£150; $150-$200

T914 Alfa Romeo car -£100-£150; $150-$200

T915-Mini-Ford car -£100-£150; $150-$200

T103-Police Patrol car -£300-£350; $450-$500

T102-Popular pedal car -£200-£250; $300-$400

T130-4-wheel tractor -£100-£150; $150-$200

T135 3 wheel tractor -£75-£100; $100-$150

T151-Demon racer go-cart -£100-£150; $150-$200

T146-Scorcher go-cart -£100-£150; $150-$200

T145-Tracmaster -£100-£150; $150-$200

T126-Tipper truck -£75-£100; $100-$150

T160-Pony cart -£200-£250; $300-$400

1972

Tri-ang Pedigree (1971) Ltd. A member of the Barclay Toy Group Ltd.

TW 3152- Rally car -£100-£150; $150-$200

TW 3144-Formula 3 racing car -£200-£250; $300-$400

TM 2020 Junior farmer -£75-£100; $100-$150

Others repeated and as priced in previous catalogue.

1973

106009 Pink Panther -£150-£200; $200-$300

106003 Chaser -£50-£75; $70-$100

106006 Charger -£100-£150; $150-$200

Others repeated and as priced in previous catalogue.

1974

101003 Strida racer -£50-£75; $70-$100

101002 Groundhog -£40-£50; $60-$70

Others repeated and as priced in previous catalogue.

1975

103002 Indianapolis Go-kart -£200-£250

103004 Silverstone Go-cart -£150-£200

Others repeated and as priced in previous catalogue.

1979

M. V. Tri-ang Ltd. (A Morris Vulcan Company)

4005 Moon/fun buggy -£50-£75; $70-$100

4015 Jeep renegade -£75-£100; $100-$150

4009 Chaser police car -£50-£75; $70-$100

4007 Fire Chief -£50-£75; $70-$100

4003 Sprite -£75-£100; $100-$150

4400 Tractor -£100-£150; $150-$200

1996-2000

Tri-ang production continues on limited all-plastic output.

Tri-ang toy supplement Catalogues missing are: 1966, 1969, 1972, 1973 through to 1990 (although many of the same pedal cars are carried through from one year to another). We wish to purchase these from readers to fill the gaps in our archives, so please kindly help! Contact the author on +44 (01273) 601960.

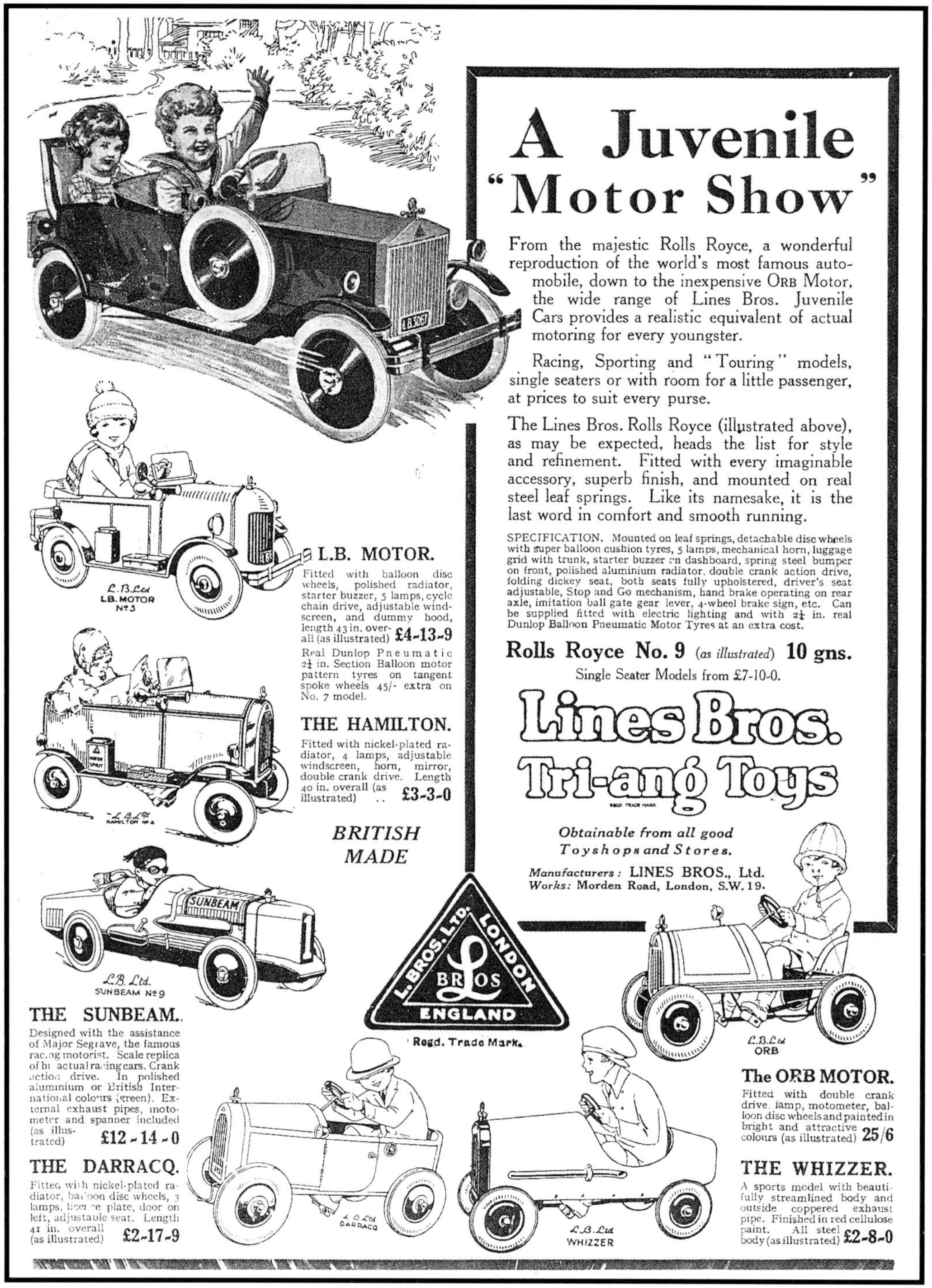

Re-printed from the *Boy's Own Paper*, issue November 1927.

Further Reading

Listed are most of the books available on pedal cars.

Bloemker, Larry; Genet, Robert; Weirick, Ed. *Pedal Cars*. WI., U.S.A. M.B.I. Publishing Co., 1999

Gardiner, Gordon and Morris, Alistair. *The Price Guide and Identification of Automobilia.* U.K. 1982; re-printed 1987.

Gardiner, Gordon and Morris, Alistair. *The All Colour Directory of Metal Toys*. U.K., Salamander,1984.

Gardiner, Gordon and O'Neil, Richard. *Toy Cars.* London, U.K. Tiger Books,1989.

Gardiner, Gordon and O'Neil, Richard. *Toy Cars*. U.K., Salamander Books, 1989; reprinted 1990.

Garton, Jane Dwyre. *Pedal Cars – Chasing the Kidillac.* PA., U.S.A. Schiffer Publishing Co., 1999

Gunnell, John A. *A Collector's Guide to Automobilia.* U.S.A., Krause Publications,1994.

Gurka, Andrew G. *Pedal Car Restoration and Price Guide.* U.S.A., Krause Publications,1996.

Herman, Paul, *Voitures a Pedales*. France, Glenat editions, 1989.

Grandvoinnet, M. *Eureka- Un Siecle de Jouets de Reve*. Paris, Librairie du Collectionneur, 1991.

Hutchinson, Ken and Johnson, Greg. *Automotive Toys 1925-1941*. Kentucky, U.S.A., Collector Books, 1997.

Levitt, Jeffrey S. *The World of Antique Toys.* London, U.K. Mint & Boxed, 1990.

L.W. Books. *Pedal Cars Vols.1 – 5.* Gas City, U.S.A., 1990-1999.

L.W. Books. *Riding Toys.* Gas City, U.S.A.,1992.

Massucci, Edoardo. *Cars for Kids.* Milano, Italy, Automobilia, 1982.

Miller's Collectors Cars Price Guide. 1991 to date (and on-going every year) Millers Publications Ltd., Kent, U.K.

New Cavendish Books. *Tri-ang Toys 1937/38.* (re-printed hardback version of catalogue along with 1939/40 supplement., London, U.K., 1988).

Pennel, Paul. *Children's Cars.* U.K., Shire Publications, 1986.

Weiner, Geoffrey, G. *Juvenile Automobiles Volume 1.* Hove, U.K. Tarik Press.1996.

Weiner, Geoffrey, G. *Juvenile Automobiles Volume 2.* Worthing, U.K. Apollo Press, 2000.

Weiner, Geoffrey, G. *Royal Children's Cars.* (In preparation now for Schiffer Publishing Co.)

Whyley, David. *Austin Pedal Cars.* U.K. Arthur Southern Limited, 1999

Wothington-Williams, Michael. *Automobilia, A Guide for Collectors*. U.K., Batsford/R.A.C.,1979.

Listed below are some of the places that have pedal cars together with vintage, veteran, and classic cars along with related automobilia, models and motoring ephemera etc.

Automobilia Motor Museum, St. Stephen, St. Austell, Cornwall, England

The National Motor Museum, Beaulieu, Hampshire, England.

The Royal Children's Cars, Sandringham Estate, Norfolk, England.

The London Toy and Model Museum, Craven Road, London, England.

The Children's Pedal Car Museum, Kingston-upon-Thames, Surrey, England. (Due to open 2002)

The Brighton Toy and Model Museum, Station Road, Brighton, Sussex, England.

The Bentley Wildfowl and Motor Museum, Halland, East Sussex, England.

The Cotswold Motor Museum, Bourton-on-the-Water, Cheltenham, Gloucester, England.

The Lakeland Motor Museum, Grange-over-Sands, Cumbria, England.

The Toy Museum, Bridge Street Row, Chester, Cheshire, England.

Bethnal Green Museum of Childhood, Cambridge Heath Road, London, England.

Haris Testverek Car Museum, Budapest, Hungary.

The Haynes Motor Museum,Yeovil, Somerset, England.

Heritage Motor Museum, Gaydon, Warwickshire, England.
Musée National de L'Automobile, Mulhouse, France
Musée de LAutomobiliste, Nice, Cannes, France.
Zentrum für Ausergewöhnliche Museen, Munich, Germany.
The Norsk Teknisk Museum, Oslo, Norway.
The Warwick Henderson Gallery, Parnell, Auckland, New Zealand.

U.S.A.

There are numerous museums and collections where you can see pedal cars in the U.S.A., too many to list here. It's best to check with the various tourist offices in the particular state or location that you intend to visit. Many weekend shows and conventions are held up and down the country that are dedicated to this interest. Again, it's best to check with the various tourist offices, local papers and magazines, and people in the know. George Medinila's website is a good first step to find contacts in the American pedal car world; log on to www.palosverdes.com/pedalcars.

There must be many other places up and down the country as well as very many more in Europe and the rest of the World. I have listed all the ones that I am aware of during the production of this book. If you have knowledge of any others, please let me know, be they small local museums, private collections on view to the public or the odd pedal car on display in your local museum or collectors shop. Many collectors' shops and stores either sell or display pedal cars, be they old, new or just relics-we want to know! There are also many collectors who may grant a personal visit from a like-minded collector and enthusiast, however please do contact them in the first instance before you ever visit!

There are also books, brochures, sales literature and other related material 'out there' that include or feature pedal cars, that are still waiting to be discovered. We dearly want to know about anything found to help with this ever-growing hobby of ours! Please share with us that old sales brochure or any other advertising material, whether it's ephemeral or in the form of cardboard or enamel signs. We would feature it in future editions of this book and credit you. Please contact the author.

Club Contacts

The Pedal Car Collectors' Club
The Secretary (P.C.C.C.)
4-4a Chapel Terrace Mews,
Kemp Town
Brighton BN2 1HU
East Sussex
England (01273) 601960

http://www.carsofbrighton.co.u.k.
http://www.brmmbrmm
http://www.eurosurf.co.u.k./Cars
http://www.Localsiteuk.com
http://www.tri-angsociety.co.uk

The Austin J40 Pedal Car Club
(Being the club for J40 and Pathfinder cars)
The Secretary (Mary Rowlands)
21, Forest Close,
Lickey End,
Bromsgrove,
B60 1JU Birmingham (U.K.) (01527) 876293
www.j40club.co.uk

e-mail: cars@kemptown-brighton.freeserve.co.u.k.
info@localsiteuk.com

If you log on to our main website you will find that we are 'linking' to other pedal car and automobilia sites around the world and hope to build up a comprehensive data base for all interested parties.

The author as a boy in his Tri-ang "Eighty" B (stock No.6001 from the 1957-1958 catalogue). Two versions were made, this the B version with both front and rear chrome bumpers with over-riders, and the A version with only a front bumper. Top right photo shows the B version fully restored. £450-£500. Note: Similar models with deluxe features are shown in the 1955 through to 1958 catalogues.